What Have We Learned About Science and Technology from the Russian Experience?

What Have We Learned About Science and Technology from the Russian Experience?

Loren R. Graham

STANFORD UNIVERSITY PRESS
STANFORD, CALIFORNIA

This book results from the Donald M. Kendall
Lecture Series in Soviet Affairs at the Center for Russian
and East European Studies, Stanford University

Published with the assistance of the Donald M. Kendall
gift to the Center for Russian and East European
Studies, Stanford University

Stanford University Press
Stanford, California

Printed in the United States of America
CIP data appear at the end of the book

To Russian scientists and engineers:
They have created, they have suffered,
and they have instructed

Acknowledgments

During the many years that I have worked on the history of Russian science and technology, I have incurred far too many and larger debts than I can ever acknowledge adequately. For support of my research in recent years I am particularly indebted to the John D. and Catherine T. MacArthur Foundation (and especially to Adele Simmons, Victor Rabinowitch, Kennette Benedict, and Andrew Kuchins among the foundation officers); the National Endowment for the Humanities (and to Daniel Jones); and the Sloan Foundation (and to Ralph Gomory and Michael Teitelbaum). When I gave the Kendall Lectures at Stanford University in 1995, upon which this book is based, I was hosted and assisted by Norman Naimark and Irina Barnes, who helped to make my visit a memorable occasion for me. During that visit many scholars asked stimulating questions and made suggestions; I remember in particular the contributions of Timothy Lenoir, Walter Vincenti, Alexander Dallin, Gail Lapidus, David Holloway, Gregory Freidin, David Tyack, and Elisabeth Hansot. Helen Tartar of the Stanford University Press attended the lectures, encouraged me to write a book based upon them, and has assisted me in many ways during the time it was being prepared. I was privileged to have colleagues who were willing to read early drafts of the book or of chapters in it and to give me their comments and critiques. They included, in the United States, Peter Galison, Douglas Weiner, Everett Mendelsohn, Evelyn Fox Keller, Carl Kaysen, Thomas Parke Hughes, Michael David-Fox, James Andrews, Linda Lubrano, Harley Balzer, and Michael Gordin. Colleagues from abroad, who discussed issues in this book with me in such distant places as Halle, Germany; Trondheim, Norway; and Moscow, Russia, included György Péteri, Irina Dezhina, Wolfgang Arnold, and Manfred Heinemann. A colleague with whom I have enjoyed discussion of Russian science on countless occasions is Paul Josephson. Vyacheslav Gerovitch is an indefatigable researcher who found many bits of information that I needed. And my assistant at MIT, Jerilyn Edmondson, has seen this manuscript through many stages and has helped me in countless ways. Lastly, my wife, Patricia Albjerg Graham, has always been the person whose advice I most valued, both as a friend and as a scholar. None of these people are of course responsible for remaining imperfections.

Loren Graham

Grand Island, Lake Superior
October 1997

Contents

Preface

By the early 1980's, the Soviet Union had the world's largest community of scientists and engineers, exceeding in number those in the United States by 10 to 30 percent, depending on the definition of degrees and fields.[1] The rapid development, in just a few decades, of such an immense scientific establishment in a social and economic environment strikingly different from the West, the birthplace of modern science, presents us with an unusual opportunity. We can better understand science and technology as social and intellectual institutions if we examine the extent to which they take on different forms in sharply contrasting environments.

This book, then, is not primarily about Russia and the Soviet Union, but about science and technology, with illustrations from the Russian experience of some of the characteristics of science and technology. These examples point to a discussion of an issue very active in the West today: whether science is a social construction. The Russian example, not commonly cited in the West, adds additional dimensions to this current controversy.

To what extent are science and technology affected by the specific environments in which they develop, and to what extent do they reflect universal concerns? Is science a social construction or is it a reflection of the natural world? Are science and technology inherently westernizing influences, or can they promote, or be used for, anti-Western purposes?

The collapse of the Soviet Union in 1991 has submitted science in Russia to new tests that can increase our understanding of science. During the existence of the Soviet Union, scientists and engineers were generously supported there financially but were sharply restricted politically and ideologically; Russian scientists and engineers after the fall of the Soviet Union have been given political freedom but have been deprived of much of their previous financial support. Science and technology, in both the Soviet and post-Soviet periods, have been submitted to quite extraordinary pressures. How robust are science and technology under stress? What is more important to science, freedom or money? Do the ways in which science and technology are formed under authoritarian and centralized controls give us insights into the relative advantages and disadvantages of democracy as an environment for technical creativity and social benefit? How welcoming were Russian scientists and engineers to democracy and a free market? How much were they willing to change their own institutions when the country in which they lived overthrew an authoritarian system that actually gave them enormous advantages and perquisites, at least compared to what followed?

Finally, the experience of Russia sheds much light on the place of technology in modern society. To what extent should technology be under the control of ordinary citizens and to what extent should it follow the preferences of engineers and technical experts? On this question, I have in the final chapter of this book compared the experiences of Russia, China, and the United States.

I have studied Russian science for many years and have written at considerable length elsewhere on all the questions raised above. The invitation to give the Donald M. Kendall Lectures at Stanford University in 1995 was the impetus to write down in a concise form the conclusions I had reached on some of these questions. The book that follows is a revised and expanded version of what originated as the Kendall Lectures, and might be described as a small book about big questions.

What Have We Learned About Science and Technology from the Russian Experience?

CHAPTER ONE

Is Science a Social Construction?

What does the Russian experience tell us about the nature of science? Let me explain why I find this question intriguing, and why I believe that the Russian experience sheds light on some of the main issues dividing historians and sociologists of science today.

In the fields of the history and sociology of science, the most striking change in recent decades has been the rise in contextualism, a growing recognition that science is embedded in society and must be studied in societal terms. In the field of the history of science, the "internalism" prevalent in the 1950's and 1960's was more and more challenged in the 1970's and 1980's by "externalism." Internalists were scholars who emphasized the power of scientific ideas and the significance of experimental findings as the major influences on the growth of scientific knowledge. Externalists, who represented a newer trend in the history of science, stressed social, economic, and other nonscientific influences on the development of science.

Among sociologists of science, a similar shift toward contextualism was growing. The older Mertonian sociology of science studied the norms of science, its reward system, and the growth and demise of disciplines and subdisciplines, but it did not study the content of science itself.[1] An assumption of the

Mertonian school was that social context may influence the careers of scientists and their institutions, but not scientific knowledge, which was treated as a "black box" and left for scientists to discuss.[2] But the new sociologists of science who began to emerge in the 1970's addressed the question of whether the very knowledge produced by scientists, including what is accepted as the best knowledge, is not shaped and formed by the society that surrounds it. According to the new approach, even the content of the densest "hard" sciences, such as physics, mathematics, and biology, can be seen as socially formed. Rooted in Peter Berger and Thomas Luckmann's 1966 book *The Social Construction of Reality*, this new trend became known as "constructivism" and was promoted in the 1970's and 1980's by sociologists of science in Britain at the universities of Edinburgh, Bath, and York.[3] At Edinburgh, in particular, Barry Barnes, David Edge, Donald MacKenzie, Steven Shapin, and Andrew Pickering were especially articulate proponents of what came to be known as the sociology of scientific knowledge, or SSK.

These representatives of new trends in the history and sociology of science were united in their belief that science is a part of culture, and that it has both natural philosophical and social dimensions. In order to test this hypothesis, one would have thought that there would be a comparison of science in contrasting cultures. Surprisingly, however, relatively little research has been done that compares the growth of scientific knowledge in societies with strikingly different cultural and political traditions. On the contrary, most of the new contextual studies have been on topics in Western science, such as the scientific revolution of the seventeenth century, or modern physics and biology in Europe and America.[4] As they sought to identify the ways in which science is a part of cultural domains, scholars were working, on the whole, within a rather narrow set of such domains, and ones that were, in many instances, a part of their own mentalities. Scholars rooted in Western traditions were studying Western science and attempting to identify the particular effects of Western society on that science. The task was not an impossible one, as is evidenced by some of the good work that has come out of it, but it was surely a difficult assignment. The problem of reflexivity here is enormous. How well can a scholar identify the influence of factors on the science that he or she studies when those same factors may be a part of his or her own analytical framework? By studying such a confined portion of the spectrum of available social contexts, and one in which they were themselves embedded, the followers of the social study of science often failed to reveal the full richness of their own intellectual program.

A useful way of testing the "social constructivist" thesis in science studies is to examine the evolution of science in a society that is distinctly different from those of Western Europe and North America, looking to see how the different environment affects science. I would like to propose Russia as a particularly appropriate case study for the examination of the social constructivist hypothesis. No one will deny that Russian society and culture have in the 1,000 years of Russian history differed from society and culture in Western Europe, where modern science was born. Russia has followed a different economic path from that of Western Europe and America, and it has religious, political, and cultural traditions quite unlike those of its Western neighbors. If the social constructivist thesis is correct, Russian science should be very different from Western science.

The most fruitful comparisons are not, however, made between entities that are totally different; rather, they emerge when one studies entities that are similar enough that some common elements can be seen but different enough that the variations can be studied. Russian science fits these criteria well. Imported initially from Western Europe, it took root and developed in distinct ways. The study of Russian science may, therefore, be a more finely tuned test of the social constructivist thesis than a study of Chinese science, since science in ancient China achieved a level of development and independence for which there is no equivalent in old Russia. Science in Russia is recognizably Western, in the sense that it was brought to Russia from Western Europe, but it is simultaneously Russian, in the sense that for almost three centuries it has continued to develop in an environment distinctly different from that of its origin. The effects of the different environment on a given scientific field may be more visible in the case of Russia than in that of China, since in the Russian case, the history of the field can often be traced back to Western European origins. Any changes that have occurred since that time are likely to be visible. In the Chinese case, the situation is more complicated because of the existence of an older native scientific tradition.

My research of Russian science as a test of the social constructivist thesis took on new significance in 1996, when Alan Sokal, a physicist at New York University, published a highly effective spoof of social constructivism in the journal *Social Text*.[5] For the first time, arguments about constructivism reached the pages of popular publications, such as *Newsweek*. In this chapter I present some of the results of my research. As a preview of my conclusions, I shall observe that in my opinion, the history of Russian science reveals both the strengths and the weaknesses of the social constructivist approach.

One of the most fruitful principles of social constructivism when applied to the history of Russian science is the principle of "symmetry." Originally posed by David Bloor as a part of his "Strong Program" in the sociology of science, the principle states that sociologists should explain what is considered to be "true" and "false" in a symmetrical fashion. Just as one may use sociological explanations to interpret false beliefs such as phrenology or mesmerism, so one should use them to explain what is currently considered true, such as the existence of atoms, electrons, and genes. As Trevor Pinch has observed, the commitment to symmetry was highly counterintuitive; within science, reliance on social factors was usually synonymous with the explanation of error.[6]

Specialists in Russian and Soviet studies have been particularly reluctant to accept the principle of symmetry, although they have often supported externalism.[7] They have recognized social factors—especially political interference—as the explanations for "bad" Russian and Soviet science, but they have not usually relied on social factors for the explanation of "good" Russian and Soviet science. When Soviet science departed most dramatically from Western science, as it did in the case of Lysenko's mistaken form of biology, they have readily explained that departure as the result of Marxist ideology and political interference. When, however, Russian science produced achievements that won recognition abroad—in mathematics, physics, biology, and psychology—they have called it "international science" and have usually refused to see the influence of any social factors that were specific to Russian culture or society. Russian-area specialists have, therefore, presented a challenge to social constructivism, maintaining that in Russia and the Soviet Union, social factors explain "errors," but not good science. Good science, by implication, is the same everywhere. The principle of symmetry, it seems, does not work for Russia.

Is this conclusion correct? Do social factors explain only the disasters of Russian science and not its triumphs? On these questions, I line up firmly with the social constructivists. It is my view that social factors are powerful explanations for the most brilliant pages in the history of Russian science, as well as for its most unfortunate episodes.

A topic within the history of Russian science on which I have applied the principle of symmetry is the role of Marxism in Soviet science. My efforts have received a mixed reception. When I have pointed to Marxism as one of the factors behind ideological repression of Soviet science, on topics such as genetics, cybernetics, relativity physics, and quantum mechanics, my efforts have generally won applause. All specialists know that Lenin wrote a book entitled *Materialism and Empiriocriticism* in which he expressed reservations about the new physics, fearing that relativity physics might be seen as reducing the significance of philosophical materialism. They also know that many Marxists in the 1930's and 1940's expressed the view that the probabilistic approach of quantum mechanics undermined the concept of causality held dear by many philosophical materialists and Marxists. And Lysenko's Lamarckian views of genetics could easily be linked to the desire for a revolutionary transformation of society embedded in Soviet Marxism. In every case in which threats to science could be identified with Marxism, the Western readers of my works generally agreed with my efforts.

However, whenever I would apply the principle of symmetry and indicate that Marxism might have something to do with a few of the brighter pages in the history of Russian science, I met resistance. I pointed out, for example, that Lev Vygotsky's work on the relationship between thought and language, Aleksandr Oparin's work on the origin of life, V. A.

Fock's work on relativity theory, and A. N. Kolmogorov's writings on the foundations of mathematics could all be connected to Marxism. All of these scientists explicitly stated that Marxism was important to their work. Furthermore, the connections that they drew to Marxism do not seem trivial but substantial. But my interpretations here were stoutly resisted by quite a few Western readers. This opposition arose from the fact that Vygotsky, Oparin, Fock, and Kolmogorov were all internationally recognized scientists. Vygotsky and Kolmogorov are two of the great names in their respective fields, psychology and mathematics, in the twentieth century. It was acceptable to link Marxism to the work of disreputable scientists like Lysenko but unacceptable to link it to eminent scientists. The principle of symmetry was not to be admitted in the history of Soviet science.

An example of the unwillingness of some reviewers to countenance the possibility that Marxism might have had an influence on Soviet scientists in their internationally recognized work was the review of one of my books by Valentin Turchin that appeared in 1988 in *Nature*.[8] Turchin evidently thought that if I believed that Marxism had occasionally had a beneficial influence on science, I was some sort of Marxist myself, and he called for a "perestroika" in my philosophical views. I happen to believe that Islam, Christianity, ancient Greek philosophy, and a variety of other philosophical and religious viewpoints also had, at moments, positive influences on science, but I do not sign up for those doctrines in saying so. We need to achieve the sophistication in our understanding of Marxism that we have achieved with these other viewpoints; historians of science are now perfectly comfortable, for example, in seeing simultaneously that dogmatic Christianity hindered the acceptance of Darwin's theory of evolution, but that a more sophisticated Christianity energized the work of Isaac Newton and Blaise Pascal. Perhaps now that Marxism has receded as a political threat, we shall be able to make this transition in understanding its sometimes contradictory effects.

Some of the best-known achievements of Russian science cannot be understood without a consideration of Marxism. In other publications, I have described this relationship more fully.[9] Here I shall cite only a few examples. One of the most interesting ones is the psychologist Lev Vygotsky.

Lev Vygotsky

In the past two decades, Lev Semenovich Vygotsky has attracted much attention in the West, especially the United States.[10] The philosopher of science Stephen Toulmin has called Vygotsky "the Mozart of psychology" and has compared him to Freud and Piaget.[11] Jerome Bruner observed that "Vygotsky was plainly a genius."[12] James Wertsch asked in 1988, "Why is it that this Soviet scholar is having so much influence on Western thought more than a half century after his death?"[13]

At the heart of Vygotsky's work on thought and language is a Marxist critique of Piaget's belief that when a child first begins to use language, the child is in an "autistic stage"—that is, the child uses language only to himself; he is thinking out loud.[14] Vygotsky found this approach unacceptable because to him it smacked of French Cartesianism, a dualistic dividing of the inner self from the external environment. Vygotsky believed that a Marxist approach to psychology demanded that these two worlds be brought into a unitary world described by historical materialism. He was explicit in tying this effort to Marxism, although the first edition of his great work *Thought and Language* published in the United States dropped this section on the grounds that Vygotsky didn't really mean it, that he was just engaging in ideological decoration.[15]

Vygotsky found a way of overcoming Piaget's dualistic worldview by centering on the moment when the child comes to realize that every object has a name. This is the moment when thought and language become socially connected. As he wrote:

> The nature of the development itself changes from biological to socio-historical. Verbal thought is not an innate natural form of behavior but is determined by a historical-cultural process and has specific properties and laws that cannot be found in natural forms of thought and speech. Once we acknowledge the historical character of verbal thought, we must consider it subject to all the premises of historical materialism. . . . The thesis that the roots of human intellect reach down into the animal realm has long been admitted by Marxism; we find its elaboration in Plekhanov [an early Russian Marxist]. Engels wrote that man and animals have all forms of intellectual activity in common; only the development level differs.[16]

Thus Vygotsky developed for the explanation of the interrelation of thought and language a scheme that contained a high degree of inner consistency and arrived eventually at Marxist conceptions of social development.

Vygotsky then took the analysis of thought and language to a higher stage, one in which higher mental functioning is heavily influenced by culturally conditioned language. As the child learns to read and becomes more sophisticated, language influences his or her thought in ever more subtle ways. Growing to adulthood, his or her mode of thinking is heavily influenced by all the literary and cultural media of the environment. A corollary of this thesis is that people in distinctly different societies and those in underdeveloped tribal ones think in different ways. In order to understand better the way in which thought and language interact in this most advanced stage of the adult in modern society, Vygotsky paid much attention to literary analysis, semiotics, and linguistics. His early work was known to the great Soviet linguist Mikhail Bakhtin, who similarly emphasized the influence of society on modes of thought. Bakhtin has also in recent years attracted much attention in the West, becoming almost a cult figure among some intellectuals.[17]

To the historian of science, showing how Vygotsky's ideas were connected to the political milieu in which he developed, the important thing is not whether his views are considered correct today. (Some psycholinguists agree with him; others do not.) Psychology and linguistics have advanced since he did his work, and it seems now to some psychologists that he underestimated the role of genetics and innate neurological structures. The important facts to the historian, however, are that Lev Vygotsky was one of the great psychologists of this century and that his views can be related to his social milieu. The social constructivist interpretation of Vygotsky is fruitful.

Many other examples exist of Russian scientists who made use of Marxism in their work in ways that attracted international attention. The British astrophysicist Stephen W. Hawking wrote in his 1988 best-selling book *A Brief History of Time* that some Soviet opposition to the "big bang" theory of the universe was linked to Marxism;[18] working in close contact with a number of these Soviet researchers, including E. Lifshitz, I. Khalatnikov, and A. Linde, Hawking came to oppose the big bang theory himself, supporting instead a version of an "inflationary model" worked out by Linde and others in the

1980s. Other Soviet scientists who connected Marxism to their work in interesting ways included the psychologists A. R. Luria,[19] S. L. Rubinshtein,[20] and A. N. Leont'ev;[21] the physiologist P. K. Anokhin;[22] the biologists A. S. Serebrovskii[23] and N. P. Dubinin;[24] the mathematicians A. D. Aleksandrov[25] and A. N. Kolmogorov;[26] the astronomer-mathematician O. Iu. Shmidt;[27] the physicists S. Iu. Semkovskii,[28] D. I. Blokhintsev,[29] and G. I. Naan;[30] and the astrophysicists V. M. Ambartsumian[31] and A. L. Zel'manov.[32]

By looking at the views on the foundations of mathematics of the eminent Soviet mathematician A. N. Kolmogorov and comparing them with prevalent views in the West at approximately the same time, we can gain an insight not only into the social construction of knowledge in the Soviet Union but also into the contemporaneous social construction of knowledge in America and the West. In the immediate pre–World War II period, the articles on "Mathematics, Foundations of" and "Mathematics, Nature of" in the *Encyclopedia Britannica* were written by two famous Cambridge scholars, Frank Plumpton Ramsey and Alfred North Whitehead. The equivalent articles in the major Soviet encyclopedia were written by Aleksei Kolmogorov, a mathematician of world rank. Let us compare the articles.

The points of difference arise on the most essential questions of mathematics: "What are the origins of mathematics?" and "What is the relationship between mathematics and the real world?" According to Kolmogorov, mathematics is "the science of quantitative relations and spatial forms of the real world."[33] It arose out of "the most elementary needs of economic life," such as counting objects, surveying land, measuring time, and building structures. In later centuries, mathematics became so abstract that its origins in the real world were sometimes forgotten by mathematicians, but Kolmogorov reminded them that "the abstractness of mathematics does not mean its divorce from material reality. In direct connection with the demands of technology and science the fund of knowledge of quantitative relations and spatial forms studied by mathematics constantly grows."[34] Kolmogorov then went on to sketch a history of mathematics in which its growth was intimately related to economic and technological demands. His views were consistent with Marxism's insistence on the material world as the source of human knowledge and on technical needs as a motivating force. Kolmogorov quoted Friedrich Engels on how mathematics was a reflection of material relationships and answered practical needs in its early history. According to Kolmogorov, despite the fact that mathematics later grew to be a highly abstract field, it never lost its organic tie to material reality.

Looking at the articles in the *Britannica* written at approximately the same time by Ramsey and Whitehead, one sees a very different analysis. According to Ramsey, mathematics is not a reflection of material relationships, but a logical system about which truth or falsity is not an important question for the mathematician. Ramsey asserted, "as a branch of mathematics, geometry has no essential reference to physical space";[35] the mathematician "regards geometry as simply tracing the consequences of certain axioms dealing with undefined terms, which are really variables in the ordinary mathematical sense, like x and y. And he demands of his axioms, not that they should be true on some particular physical interpretation of the undefined terms, but merely that they should be *consistent* with one another."[36] In a somewhat similar vein, Whitehead defined mathematics as the "science concerned with the logical deduction of consequences from the general premises of all real reasoning," with no reference to the influence of the material world. In fact, Whitehead maintained that "the act of counting" is "irrelevant to the idea of number."[37]

Clearly, Kolmogorov was influenced by Marxism in his views on the development of mathematics. In that sense, his article is an example of the social construction of knowledge. But is it not also clear that Ramsey's and Whitehead's articles also bore the imprint of the time of their authorship, the pre–World War II period? Stephen Toulmin, a philosopher of science, was educated in England at this time, and he described the prevalent attitudes of his teachers: "As recently as the 1930s, when I first acquired my ideas about 'science,' the most characteristic mark of the scientific attitude and the scientific task was to select as one's preferred center of attention the purest, the most intellectual, the most autonomous, and the least ethically implicated extreme."[38] He went on to describe "the factual, unemotional, antiphilosophical, class-structured, and role-oriented attitudes of the English professional classes between the two world wars."[39] Do not these attitudes show up in Ramsey's and Whitehead's discussions of mathematics? And are we not justified in thinking that *both* the articles by Kolmogorov and the ones by Ramsey and Whitehead display signs of social construction?

In the case of the Soviet Union, the social construction of knowledge was not left to accident, but was aided by a state system of education that required every university student to take courses in dialectical materialism, the Marxist philosophy of science. The most general principles of this philosophy were the following:

1. The world is material and is made up of what current science would describe as matter-energy.
2. The material world forms an interconnected whole.

3. Human knowledge is derived from objectively existing reality, both natural and social; being determines consciousness.

4. The world is constantly changing, and, indeed, there are no truly static entities in the world.

5. The changes in matter occur in accordance with certain overall regularities or laws.

6. The laws of the development of matter exist on different levels corresponding to the different subject matters of the sciences, and therefore one should not expect in every case to be able to explain complex biological and psychological phenomena in terms of the most elementary physicochemical laws.

7. Matter is infinite in its properties, and human knowledge will therefore never be complete.

8. The motion present in the world is explained by internal factors, and no external mover is therefore needed.

9. The knowledge of human beings grows with time, as is illustrated by their increasing success in applying it to practice, but this growth occurs through the accumulation of relative—not absolute—truths.[40]

This philosophy of science is actually a quite sensible one and corresponds to the implicit views of many working scientists all over the world. The most common philosophical viewpoint of these scientists is "realism," which means that they believe in the existence of an objectively existing real world. Dialectical materialism is highly compatible with this viewpoint, although its suspicion of reductionism (the view that all phenomena can be described in terms of the principles of physics, if knowledge were only complete) is sometimes resisted by physicists. On the other hand, many biologists are in agreement with the antireductionism of dialectical materialism.

But dialectical materialism tries to steer a middle path between reductionism and vitalism.

Not surprisingly, this philosophy seems to have left a lasting impact on Russian students. In 1996, five years after the collapse of the Soviet Union, G. B. Zhdanov, a Russian physicist at the Lebedev Institute of Physics (FIAN), one of the most famous and distinguished science institutions in Russia, polled 60 of his colleagues on their attitude toward the old official Soviet philosophy of science, dialectical materialism. He asked them to select one of the five following statements as best describing their philosophical views:

1. Dialectical materialism is the only true philosophical orientation.
2. Dialectical materialism is better and preferable to all others, but needs to be further developed in an essential way.
3. Dialectical materialism has just as much right to existence as do other scientific orientations.
4. Dialectical materialism, having compromised itself in the recent past in our country with unsuccessful evaluations of a series of scientific developments, must be considered a dead end.
5. Dialectical materialism is not interesting to me.[41]

According to Zhdanov, 13.4 percent of his respondents chose sentence no. 1; 36.8 percent chose sentence no. 2; 21.6 percent chose sentence no. 3; 11.4 percent chose sentence no. 4; and 15.5 percent chose sentence no. 5.

If we summarize the results of this poll, we see that approximately half of the respondents at the Lebedev Institute of Physics in 1996 rated dialectical materialism above all other philosophies of science, and over 70 percent thought that it was at least as valuable as other philosophies of science. This finding will surprise many Western observers, among whom dialectical materialism does not have a high reputation. Zhdanov's methodology was not impeccable, and it is possible that his poll exaggerates the popularity of dialectical materialism among Russian physicists. How representative was the sample? Did Zhdanov, a senior researcher, tend to ask colleagues of similar age, people who had been thoroughly shaped by the Soviet experience, and ignore younger scientists? It may be hoped that further research, with a more explicit methodology, will help to answer these questions. But Zhdanov's poll, even if its results are somewhat skewed, shows the flaws in the old view that Marxism was forced upon unwilling natural scientists by Marxist philosophers in the Soviet Union. Even today, long after the demise of the Soviet Union, there are scientists in Russia who continue to write on what they see as the useful insights brought to science by Marxism.[42] At the moment, with the Soviet Union gone, there may actually be more support for dialectical materialism among scientists in Russia than there is among professional philosophers.

Thus, the history of Russian science, which contains so many paradoxes, presents one more: the aspect of Soviet Marxism that is usually considered the weakest by Western observers—the philosophy of nature known as dialectical materialism—was actually the strongest, and the aspect of Soviet Marxism that still attracts much attention in the West—the theory of economics and history presented by historical materialism—was actually the weakest. Historical materialism was based on the view that the proletariat, growing ever stronger and more powerful in advanced industrial nations, would create a world revolution. Who believes this anymore? The proletariat is diminishing in strength in all advanced industrial nations, and a Marxist world revolution is increasingly remote. But the belief that nature objectively exists and that its study through science provides the most reliable information we possess continues to gain strength. Soviet scientists went further toward building an explicit and developed natural philosophy based on this view than any other group in modern history. In several of my other books, I have described in considerable detail the instantiation of this natural philosophy in Russia in the major scientific disciplines—physics, chemistry, astronomy, computer science, biology, and psychology.[43]

At this point, many of my readers may be asking, "But did not dialectical materialism have disastrous effects on Soviet science? What about Lysenkoism, the false doctrine of genetics preached for decades in the Soviet Union? Is this not an example of both the damage done to Soviet science by Marxism and of the dangers of the social construction of science?" Lysenkoism is, indeed, an example pointing to the dangers of dogmatic ideology and an extreme social constructivist view of knowledge. Let us look at it more closely.

Lysenkoism

The advent of Lysenkoism was a formidable exercise in the social construction of science. It is inconceivable that Lysenkoism could have come to the Soviet Union in the form and in the way that it did without the particular characteristics of intellectual, social, and economic life that existed there in the 1930's and 1940's. Those characteristics included a striving for transformation brought by the Revolution, a monopoly over intellectual life exerted by the Communist Party, and a deep agricultural crisis in the wake of the disastrous collectivization program.

Most interpreters of Lysenkoism in the West have emphasized the importance of a desire to transform society as an explanation for the Soviet embracing of Lysenkoism. Lysenko supported a form of Lamarckism, the doctrine of the inheritance of acquired characteristics. If people can inherit improvements acquired from the social environment, so the argument goes, then revolutionary changes in society can quickly result in the improvement of human beings. Therefore, Lysenko's Lamarckist view of heredity extended hope to the revolutionary leaders of the Soviet Union that backward peasants could be converted in a few generations into outstanding citizens who were both environmentally and genetically transformed.

This explanation of Lysenkoism is not accurate. Lysenko never claimed that his views on heredity were applicable to human beings. Indeed, he castigated eugenics and all other attempts to alter human heredity as examples of bourgeois influence on science. All the studies of the history of Lysenkoism that have been written during the past thirty years—especially those of David Joravsky and Zhores Medvedev and my own—agree that Lysenkoism was not based on human genetics.[44]

However, Lysenkoism *was* based on a hope for a great transformation, a revolutionary change in heredity, but directed not toward human beings but toward agricultural products. And here the importance of collectivization emerges as another important ingredient in the social construction of Lysenkoism.

Lysenko was conducting his fieldwork at a time when Soviet agriculture was in crisis as a result of the recent massive collectivization, and this desperate moment presented him with extraordinary opportunities to win the authorities' attention with his alleged solutions to agricultural problems. The collectivization program had been incredibly violent, involving the deportation and eventual deaths in camps of hundreds of thousands of peasants. A famine followed in Ukraine that resulted in the deaths of millions.[45]

At the time when Lysenko began his campaign for a socialist agriculture in the 1930's, there were few agricultural specialists who were willing to work energetically for the success of the new and troubled collective farms. Many agronomists of the time had been educated before the Revolution; even among the younger ones with Soviet educations, many disagreed with the collectivization policies, seeing the damage that had been done in the countryside. Among the biologists in the leading universities and research institutes, the most exciting topic of the time was not agriculture but the new genetics arising out of research on the fruit fly *Drosophila melanogaster.* Only later would it become obvious that this research had great agricultural value, producing many agricultural innovations, such as hybrid corn. In the late 1920's and early 1930's, it was easy for radical critics like Lysenko to castigate the theoretical biologists as they bent over trays of fruit flies in their laboratories at a time when famine stalked the countryside. Lysenko was a master at propaganda directed against these academic biologists; he called them "fly-lovers and people-haters."[46] Since many of the professional biologists had bourgeois backgrounds, their political loyalties were always suspect to the regime. The unwillingness of many theoretical biologists to work directly on agricultural problems was seen by the radicals as purposeful "wrecking," an effort to disable the Soviet economy and cause it to fail, rather than the result of the common division the world over between theoretical and applied biology. And the higher status claimed by many biologists for theoretical as opposed to applied investigations exacerbated the issue.

Lysenko was strikingly different from the majority of biologists and agronomists. He came from a peasant family, he was a vociferous champion of the Soviet regime and its agricultural policies, and he offered his services to agricultural administrators. Whenever the Party announced plans to cultivate a new area or plant a new crop, Lysenko came up with practical suggestions on how to proceed with the plan. He developed his various nostrums so rapidly—from cold treatment of grain, to plucking leaves from cotton plants, to removing the anthers from spikes of wheat, to cluster planting of trees, to unusual fertilizer mixes, to methods of breeding cows—that before the academic biologists could show that one was valueless or harmful, Lysenko was off announcing another technique. The newspapers invariably applauded Lysenko's efforts and questioned the motives and political backgrounds of his critics. In this environment, a peasant agronomist who promised a revolution in agriculture had enormous political advantages over sober academic geneticists who—in a time of crisis—appeared to be restraining progress by crying "not so fast," or "inadequate verification."

Completely aside from the poverty of his genetic views, Lysenko's work had significant psychological value. The primary question of the times was not so much whether his biological theories would work as whether the peasants would work. Still alienated by the collectivization program, the peasants at first found difficulty seeing very much "new" about "socialist agriculture" except the fact of dispossession. Lysenko and his followers introduced much that was new, and they worked side by side with the peasants. Every peasant who participated in Lysenko's projects was enrolling in the Great Soviet Experiment; a peasant who at Lysenko's urging planted wheat had clearly graduated from the stage when he destroyed his wheat so that the Soviet government would not receive it.

Every one of Lysenko's projects was surrounded with the rhetoric of socialist agriculture, and those who liked his projects committed themselves to that cause.

Using methods such as these, Lysenko won enormous support in political circles, among communist journalists, and among the nonscientific leaders of the educational and agricultural establishments. The fact that Lysenko was simultaneously denying the existence of the gene, that he was discarding all of modern genetics, meant less to these people than the fact that he was actually getting Soviet peasants to work in the fields and that crops were being harvested. The academic scientists could not, at that time, point to any such concrete and immediate results or benefits to society from their work.

Here, then, is an instance when social and political influences were having a massive impact on science. The social construction of science was proceeding at a breathtaking pace.

As you can see, it is my opinion that the social constructivist approach is very fruitful for the study of the history of Russian science. However, the application of social constructivism to Russian science not only demonstrates some of the strengths of that approach, but also some of the weaknesses. Since social constructivism is based on the principle of finding out how the development of science is influenced by society, in a stable society, it is better suited for explaining scientific continuities than dramatic scientific discontinuities. How well does social constructivism explain sudden changes in scientific conceptions if the society in which that change has occurred does not itself change in a fundamental way? Is it possible that in such instances, the main agents of change may not be social or cultural but instead cognitive or empirical? Social constructivism can help us to understand why Lysenkoism lasted so long, for over

30 years, showing the manifold ways in which, politically and ideologically, Lysenkoism fitted with and derived strength from Soviet society. Social constructivism is less helpful in explaining why Lysenkoism was eventually abandoned. The rejection of Lysenkoism and the embracing of Western-style genetics in the Soviet Union in the 1960's was a disruptive discontinuity that violated Soviet ideological principles that had been developed over a period of several decades and was therefore resisted by the political establishment. The Communist Party apparatus, the education ministries, and the agricultural research institutions were filled with officials who had made their careers supporting Lysenkoism and who firmly resisted any attacks upon it. The forces responsible for the social construction of Lysenkoism were still very powerful in the early 1960's, when Lysenkoism came under serious attack. That attack was at first advanced by a minority who had little administrative, political, or social clout. In the face of such strong social forces favoring Lysenkoism, how did the critics eventually succeed? An adequate explanation of how this success occurred must devote more attention to factors social constructivists often ignore: the relative strengths of differing internalist arguments about nature, and the relative success of differing agricultural practices. To put the matter more simply: the classical genetics practiced in the West was based on much more convincing scientific evidence and more tightly argued theory than Lysenkoism, and the agricultural applications that resulted from classical genetics were over time outperforming the agricultural applications promoted by the Lysenkoites in the Soviet Union. In other words, the socially constructed doctrine of Lysenkoism was being undermined by contradictory scientific evidence, a powerful alternative cognitive scheme, and the

convincing results of agricultural practices based on Western-style genetics.

Lysenkoism was constructed in the Soviet Union before the revolution of modern genetics had occurred in Western society. In the 1930's, when Lysenko built his power, the role and structure of DNA were not yet understood. Although hybrid seed corn and a few other products of genetic breeding were beginning to be applied to agriculture in the West, they were not yet well known. In other words, it was possible in the 1930's to deny the existence of the gene and still pursue agriculture. It was even possible to doubt the existence of a specific hereditary substance and be a scientist. Among Western geneticists in the 1930's, there was still uncertainty over heredity; some scientists believed that protein was the carrier of heredity, while other were beginning to suspect that DNA was the important substance.

By the early 1960's, when Lysenkoism began to come under heavy criticism in the Soviet Union, all this had changed. In 1953, James Watson and Francis Crick did the work that won them the Nobel Prize and revolutionized biology: they described the structure of DNA. By the 1950's and 1960's, the practice of agriculture was being transformed by modern genetics. When Nikita Khrushchev visited Roswell Garst's corn farm in Iowa in the early 1960's, he was staggered to learn that no farmers in Iowa still grew their own seed corn; instead, they all purchased their seed corn from commercial breeders, who used the principles of Mendelian genetics, which Lysenko denied.[47]

Lysenkoism was being threatened by scientific developments and agricultural practices in the West, which had become so cognitively convincing and so agriculturally effective that they could no longer be ignored. It would, in my view, be accurate to describe what was happening as "the obtruding of reality into the social construction of Lysenkoism." A massive social construction was being broken up by scientific evidence in favor of a material carrier of heredity—DNA—and equally massive agricultural evidence in favor of the Western form of Mendelian genetics. All of Lysenko's denials of the existence of the gene and his criticisms of Mendelian genetics could not stand up to this evidence. A social construction was being overturned.

I realize that quite a few historians and sociologists of science will say that Western genetics and agricultural practices were also social constructions at this time, and that what was happening is not best described as the refutation of Lysenko's genetics by scientific and agricultural reality, but the substitution of one social construction of genetics by another—the replacement of Lysenko's genetics by Western genetics, both of which are social constructions. In my opinion, such a criticism is valid, but very limited in its scope and misleading in its direction. Of course, all science is a social construction in the sense that it is created by members of society; it cannot be otherwise. But to insist on social construction (as it is usually understood) as the key to what was going on when Lysenkoism was overthrown by Western genetics is to miss the most important elements—the roles of cognitive factors, theoretical rigor, and practical success. I certainly do not see Western genetics at the time it displaced Lysenkoism—or at any time—as representing ABSOLUTE TRUTH, devoid of elements of social construction. We all know that genetics will continue to change, and we all know that social influences have played and will continue to play important roles in the construction and development of genetics. Recent research on modern Western genetics is showing us the importance of social influences in its formation.[48] However, the case of the displacement of Lysenkoism by Western genetics should make clear that when we analyze the formation of ideas about genetics, we should include, along with social factors, scientific evidence and cognitive rigor. Experimental evidence and straight thinking *do matter*, and Lysenko was deficient in both.

Those extreme social constructivists who refuse to admit terms like *reality* into their discourse can always find another purely social constructivist explanation for whatever contrary evidence is presented to them. Since all science is created by human beings, who are inherently social, and since, as Ludwik Fleck showed years ago,[49] even scientific "facts" can be socially selected and formed, no logical "proof" can ever be found that destroys social constructivism, just as no logical proof can refute epistemological idealism or solipsism. But after a while the arguments of the extreme social constructivists become sophistic, possibly still barely plausible from a logical standpoint, but certainly not persuasive as an explanation for the world in which we live. In the arguments of extreme social constructivists, nature tends to evaporate entirely and only social discourse remains. For this reason, some of their writings are scientifically ridiculous, as Alan Sokal demonstrated in his spoof of their views, accepted for publication by social constructivist editors.

One of the reasons why Western genetics displaced Lysenko's genetics is that it possessed much greater scientific rigor, was based on an immensely richer body of empirical evidence, and resulted in much more effective agricultural practices than Lysenko's genetics. To me, this amounts to the intrusion of reality into Lysenko's socially constructed worldview.

Although I obviously think that the arguments that many

social constructivists normally use to describe how scientific knowledge is formed are inadequate for an explanation of the overthrow of Lysenkoism, I do not believe that the Lysenko example invalidates all forms of social constructivism. I have gained much from social constructivism, although I discipline it by including empirical evidence and scientific rigor as factors crucial to the formation of science.

Some readers of my analysis of the role of social factors, including political ideology, on the development of scientific knowledge may conclude that I locate science and society on different planes, which can interact, but that I think science would be better off if such interaction did not occur. They may think that I see social factors as magnets able to deflect science from its proper autonomous course.[50] Thus, Lysenkoism was created by a powerful magnet called "Soviet ideology," but it was later discredited by objective science, free of the influence of all such magnets. In this reading, the social and cultural enter science at the moment of creation (context of discovery) but not at the moment of testing (context of justification.)

Such a reading of my views would be erroneous. It assumes that I think that science can and should exist outside of any social context. At bottom, such an interpretation on my part would be based on a Cartesian dualism in which mind can exist in an unembodied, uncaused, and autonomous state. The reality is that science is *always* both created and tested within a cultural context; science is a part of culture, not something separate from it. To hope for a science that is free from social influence is to hope for something that never was and never can be. Such an uninfluenced science is a contradiction in terms, because science is created by human beings living in society.

Science is born, developed, and tested within a social matrix, and it shows this life history in a multitude of ways.

But just because science always has a social dimension as well as a natural philosophical one does not mean that we should consider science identical to all other cultural products, such as literature, theology, and philosophy. Science is the most reliable form of knowledge that we possess, and the reason for this reliability is its relationship to reality, both social and natural (indirect and mediated though that relationship usually is). Reality is our friend, because it gives us a means of reining in our hypotheses, abandoning those that are not supported by our attempts to verify them. Lysenkoism fell short when checked in this way, and it was eventually abandoned even by the Soviet authorities, who had invested heavily in it. Some observers of this episode in the history of science, thinking in rather simple terms, will say that this process can be described as the replacement of "falsehood" by "truth," and that when Western genetics replaced Lysenkoism in the Soviet Union, science was rescued from the influence of political factors. A more sophisticated person with a better understanding of the history of science would say that this process can be described, at best, as the emergence of "relative truth," and that Western genetics is also a social product that exhibits the characteristics of its cultural evolution. But the fact that science, including the best science that we possess, is a social and cultural product should not prevent us from seeing the superior value of knowledge that stands up to empirical tests and intellectual analysis. Western genetics and Lysenkoism differed sharply in this regard.

In *Knowledge and Social Imagery*, David Bloor observed that to claim that knowledge depended exclusively on social variables would be "absurd."[51] This observation opens the door for a much more subtle and fruitful form of social constructivism than both its critics and its supporters usually grant. The factors that go into this more sophisticated form of social constructivism include not only the social, political, and ideological ones that social constructivists traditionally emphasize, but also arguments about scientific evidence and theory, as well as considerations of practical results from applications. The proportions of the various factors in this multifactorial approach will vary from case to case and cannot always be sorted out. I think that a strong argument can be made, however, that in the Lysenko affair, social and political factors played a striking role in its original rise, and cognitive and practical arguments played a strong role in its demise. But, of course, in both phases, the social construction of science was proceeding.

Conclusions

As we look back over the history of Russian and Soviet science, thinking about what it might tell us about history of science in general, a double paradox emerges: the episode in Soviet science that is usually thought of as providing a graphic example of social constructivism—the Lysenko Affair—actually provides rather convincing evidence of the weakness of the extreme form of a social constructivist understanding of science. After all, Lysenkoism not only arose in the Soviet Union, it also eventually collapsed. An adequate explanation of the Lysenko Affair must explain both events. Social factors are more powerful as an explanation of why Lysenkoism arose than it is an explanation of why it finally disappeared. On the other hand, some of the episodes in Soviet science in which social constructivism is almost never invoked—namely, those moments when achievements there were internationally recognized—are actually episodes where the power of social constructivism is evident. The strengths of some of the great schools of Soviet science—such as the Vygotsky-Luria-Leontiev school in psychology, or the Vladimir Fock school in relativistic physics, or the A. D. Aleksandrov school in geometry, or the Luzin school of mathematical functions, or the Vernadsky school in geochemistry—cannot be adequately explained without reference to social and political factors. Marxism was only one, but certainly an important one, among these factors. Anti-Marxism, hidden opposition to Soviet policies, and religion also played roles. As Douglas Weiner has shown, resistance to Soviet industrial development by ecologists strengthened their commitment to the theoretical concept of "biocenosis," the self-enclosed ecological community.[52] Karl Hall is showing us how Stalin's emphasis on engineering practice influenced the work and writings of famous Soviet theoretical physicists, such as Fock, Tamm, and Landau.[53] Religion was important among some Russian scientists, too, especially several leaders of the "Luzitanians," the followers of the great functionalist in mathematics, N. N. Luzin.[54] Luzin carried on an extensive correspondence in the 1920's with Pavel Florenskii, a Russian Orthodox priest and mathematician, and Luzin believed in an intimate connection between his intuitive approach to mathematics and religion. This intuitive approach has been one of the hallmarks of the Moscow school of mathematical functions, a significant movement in mathematics in this century. In recent years, there has been an outburst of writing in the former Soviet Union about the relationship of science to religion, just as several decades earlier there was much writing on science and Marxism. A thorough exploration of the social dimensions of Russian science must include *all* social factors, of which Marxism and religion are important examples.

Out of all this it emerges that the principle of symmetry *does* apply to the history of Russian science. Social factors were important not only in the development of Russian approaches to nature that were later judged to be unsuccessful, such as Lysenkoism, but in the development of approaches and schools from which the rest of the world benefited, such as the Vygotsky school of psychology and the Moscow school of mathematical functions. But while social factors, as normally understood, can tell us much about the history of Russian science—just as they can tell us much about the history of science in general—they cannot tell us everything. We must allow room for scientific evidence, theoretical coherence, and technological success as factors affecting the development of science. If we do not, we shall have trouble explaining why the history of Lysenkoism contains two signal events: its rise and its fall. But even more important, we do not understand science.

In this chapter I have concentrated, not on the nature of the Soviet Union and Soviet history, but instead on science and the history of science. Now I would like to apply some of what I have said to the Soviet Union itself. Just as the Lysenkoist description of biology fell short of an adequate explanation of natural reality and failed to produce successful applications in agriculture, so also the Soviet Marxist explanation of society fell short of an adequate explanation of economic reality and failed to produce the prosperous society that had been promised. The revolutionary effort to build a totally new society in the Soviet Union, one different from and superior to all others in the world, was a grand effort in social constructivism, probably the most ambitious such effort in history. In a fashion similar to the way in which Lysenko and his supporters denied the existence of the gene, Marxist economists and their supporters denied the existence of the market. Despite all the social constructivist support that Soviet theorists and politicians tried to give to Lysenkoism and Marxist economics, both have fallen into eclipse. Both the gene and the market have reemerged, and, one is tempted to add, "with a vengeance." Natural and economic realities have obtruded. I caution against drawing the conclusion that we should abandon social constructivism and begin speaking of our cognitive schemes as descriptions of objective reality, however, although some critics of the social construction of knowledge will wish to draw that conclusion. We have learned far too much from social constructivism to return to a form of naive realism. Although our present conceptions of heredity and of economic theory are vastly more effective than Lysenkoism or the Soviet form of economic Marxism, we have no justification for believing that they represent objective reality. On the contrary, we have considerable grounds for assuming that our current conceptions about heredity and economics will undergo substantial revisions in the future. More to the point, we have reason to believe that our present theories about heredity and about economics are social constructions containing elements not only of experimental evidence but of social and political influence. The challenge of the next generation is to find explanations for the growth of scientific and social knowledge that build on David Bloor's denial that science is "purely social" or "merely social."[55] The history of Soviet science and the history of the Soviet Union itself are powerful indicators that reality *does* matter.

CHAPTER TWO

Are Science and Technology Westernizing Influences?

When historians, political scientists, and other specialists in the social sciences and humanities discuss the relationship of Russia to the West, they rarely talk about science and technology. Science and technology are largely absent as subjects of research in our major centers of Russian studies, even though a strong case can be made that the history of science and technology is as important for an understanding of Russian history as the history of Russian literature, a subject considered mandatory in many Russian studies programs. In this essay, I would like, in a small way, to attempt to redress this situation. I shall do so by making some strong claims. My strongest claim will be that the course of Soviet history and the failure of the Soviet experiment cannot be understood without including science and technology as major causal factors.

The most important effect of science and technology in Soviet history has been their eroding influence on the exceptionalism that Soviet leaders originally claimed for their society. In other words, science and technology go to the heart of one of the oldest issues in Russian studies: the relationship of Russia to the West. The thesis of this chapter is that science and technology have acted powerfully as moderating influences, as forces pulling Russia toward the West, as factors reducing the differences between Russia and the West. In the latter part of this chapter, however, I shall point to recent indications of a halt, if not a reversal, of this process.

Before proceeding to the details of my argument, I must issue a few caveats. When I speak of science and technology as "causal factors" or "forces," I know that I am opening myself up to the criticism that I am a "technological determinist," a person who sees science and technology as forces shaping social development without regard to the characteristics of the society in which science and technology are situated. On the contrary, the interpretive model of historical development that I follow is not that of "determinism" but that of "interactionism." By *interactionism*, I mean that science and technology do not act alone as historical forces but mix strongly with existing social and political factors to produce results that are not explicable in either purely social or purely technical terms. It is, in my opinion, senseless to talk about the "impact" of a given technological product or scientific theory on a given society without paying equal attention to technical and social characteristics. In sum, I am neither a "technological determinist" nor a "social determinist," but a historian who maintains that science and technology are major factors—but certainly not the only ones—influencing the course of history. However, in these few pages, I emphasize the formative influence of science and technology on Russian society. If this focus makes me sound like a technological determinist, so be it. I shall try to refute the charge at another time and place, if need be.

In my opinion, the influence of science and technology on Russian and Soviet society has been so immense that a large book, and indeed many books, would be required in order to describe and analyze it in anything approaching a satisfactory way. Since I obviously do not have room to do so here, I shall merely make some broad claims and back them up to the degree that space allows.

Concentrating on the Soviet period, I see the pervasive influence of science and technology on many aspects of the Soviet cultural and political system. In abbreviated fashion, I shall sketch out specific examples of this influence in five broad areas—scholarship, information media, foreign policy, economic development, and domestic politics—giving a paragraph or two to each area.

In *scholarship*, science was a contradiction to the "revolutionary exceptionalism" claimed by the leaders of the USSR. The revolutionaries who established the Soviet regime believed that they were not only opening a new epoch in human history but also creating a new intellectual world. Once the transformation to a socialist economy was accomplished, the whole world of scholarship would also be reformed, with Marxist approaches to social and natural reality supplanting bourgeois ones. From the first years of the regime, however, natural science presented a special problem. Soviet literature and art might be different from bourgeois forms, but what about Soviet physics or biology? Answering this question proved to be far more difficult than a Western observer might suspect. The answer came with the crash of Lysenkoism.

The significance of Lysenko in Soviet biology can be summarized as follows: his reign, from roughly 1948 to 1965, was the apogee of the Soviet claim to exceptionalism; the total failure of Lysenko's form of biology in competition with the molecular biology sweeping the rest of the world was an early indicator of the fallibility of the Communist Party and the falseness of the Soviet leaders' claim that the society they were creating would be not only different from the capitalist world but superior to it. A number of later Soviet dissidents, including

Zhores Medvedev and Andrei Sakharov, took some of their earliest steps in defiance of the regime over the issue of Lysenkoism.[1] But the downfall of Lysenkoism was far from merely a school for later dissidents. Lysenko denied the reality of the gene, and the Party sanctioned this denial. Was it possible that the Party was similarly wrong in some of its other denials? As noted in chapter 1, Marxist economists similarly denied the reality of the market, and the Party sanctioned this denial also. By puncturing the first gigantic hole in the façade of official Marxist scholarship, the geneticists who overthrew Lysenkoism threw doubt on all of Marxist ideology. The result of their efforts was to pull Soviet intellectual life closer to Western intellectual life. Many Soviet intellectuals in other areas—from economics, to sociology, to philosophy, to history, to politics—soon followed suit. Soviet exceptionalism was being reduced at a rapid rate. The incontrovertible nature of the geneticists' victory over the Lysenkoites was a stunning and transforming moment in the history of the Soviet Union.

In *information media*, the emergence of new technologies of communication, which were nearly uncontrollable, caused the Soviet security organs to loosen and eventually abandon their censorship controls. It is difficult now for newcomers to the field of Russian studies to imagine how pervasive and successful Soviet control over information was in the 1950's and early 1960's. When I went to the Soviet Union as a graduate student in 1960, obtaining information from the outside world was, for the average Soviet citizen, nearly impossible. No foreign newspapers, except for those of the Western communist parties, were for sale at the newsstands; to make a long-distance call, even to another city in the Soviet Union, not to speak of a call to a foreign country, it was necessary to go to the central post office in downtown Moscow; all short-wave foreign broadcasts were jammed, and the jamming was fairly successful in the major cities; photocopiers did not exist, and even in later years when they did, they were locked up and controlled by security agents; tape recorders were not available, and adventurous souls wanting to make surreptitious recordings went to elaborate efforts to make phonograph records on used X-ray plates.

After the early 1960's the Soviet Union was flooded with wave after wave of new information technologies that replaced the vertical lines of communication supplied by official and centralized radio and television stations with the horizontal lines of communication coming from control over new communications technologies by private citizens.[2] First came the tape recorder, then a proliferation of transistor-based, small short-wave radios that could be transported easily to places where jamming was incomplete, then came photocopiers, then videocassette recorders, then computers, then modems and the whole infrastructure of the information revolution. The formerly closed society of the Soviet Union was perforated by hundreds of thousands of holes, through which outside information streamed inward and inside information streamed outward.

It is now difficult to believe, but in 1960 when I first went to the Soviet Union, if a scientist in, say, the Lebedev Institute of Physics were arrested, it might be months before we had reliable information in the West on the event. When the internationally known geneticist Nikolai Vavilov was arrested in 1940, it was several decades before we learned the details. Today, such an arrest would be known in the West within hours. New information technologies are not the only reasons for this dramatic change, but they are crucial for understanding it.

In *foreign policy,* the original assumption of Soviet leaders was that the antagonism between socialism and capitalism was so intense that cooperation between the two systems could only be episodic, used for temporary strategic purposes. Eventually one of the two world systems would triumph, and Marxism left no doubt about which one it would be. Lenin's position was that wars are inevitable among capitalist powers and that when they occur, socialists should try to convert them into class wars.[3] Early Soviet leaders believed that international conflict would bring about the emergence of successful socialist revolutions. The Soviet Union was born, after all, amid the destruction and violence of World War I, and the entire East European bloc of socialist countries arose in the aftermath of World War II. For decades, Soviet foreign policy saw conflict, both among nations and among classes, as the seedbed of positive social change.

Beginning with Nikita Khrushchev, however, a different principle based on some degree of long-term cooperation began to emerge implicitly in the speeches of Soviet leaders. At the Twentieth Congress in 1956, Khrushchev explicitly declared invalid the Marxist-Leninist principle that says that while imperialism exists, wars are inevitable.[4] What caused this momentous shift? All historians know that major events usually have multiple causes, and no doubt this shift had divergent roots as well. But Khrushchev made clear that a major factor in his thinking was a development in technology, the advent of intercontinentally deliverable nuclear weapons. In announcing his new policy of "peaceful coexistence," Khrushchev observed that

> it would be too late to discuss what peaceful coexistence means when the talking will be done by such frightful methods of destruction as atomic and hydrogen bombs, as ballistic rockets which are practically impossible to locate and which are capable of delivering nuclear warheads to any part of the globe. To disregard

> this is to shut one's eyes, stop one's ears, and bury one's head as the ostrich does when in danger.[5]

Taking notice of Khrushchev's shift in policy, U.S. and Soviet scientists began meeting together to discuss means to control nuclear weapons, at first at the Pugwash conferences and later in the U.S.-Soviet Study Group on Arms Control and Disarmament. A leading figure on the American side, the Harvard biochemist Paul Doty, later described how early contacts between U.S. and Soviet scientists on technical matters created a level of trust that enabled them to proceed subsequently to difficult political subjects such as arms control. These conversations laid the intellectual and political foundations for the policy of cooperative regulation of armaments between the United States and the USSR that were codified in the 1972 SALT I and ABM treaties.[6]

Against this background, we are justified in seeing the development of military technology and contacts between Soviet and American scientists as major factors in the most momentous change in foreign policy in the history of the Soviet Union.

In *economic development,* the desire of the Soviet Union to keep up with the pace of technical achievement in the capitalist world forced it to adopt Western designs and cooperate with Western countries in ways that had earlier been ideologically repugnant. One of the most dramatic examples here is in the field of computers. The history of computer development in the Soviet Union can be described as an early attempt to sustain Soviet exceptionalism, the belief that the Soviet Union would go on its own independent and superior path. Russia and the Soviet Union had a brilliant tradition in mathematics and the development of the algorithms that form the heart of computer science; among the early Russian pioneers were N. M. Krylov, A. N. Kolmogorov, N. N. Bogoliubov, and S. A. Lebedev. At first it looked as if the Soviet Union would be a leader in computer development. In 1950, the Soviet computer designer S. A. Lebedev produced the MESM, the first electronic, stored-program, digital computer in continental Europe.[7] Only the U.S. and the British efforts were earlier, and at that moment it would be accurate to say that the Soviet Union was at a world level in the development of computers. The MESM was developed totally independently of Western efforts and possessed its own unique architecture. In the 1960's, the Soviet Union developed and produced about 250 second-generation digital computers, the BESM-6. Although the BESM-6 was a good computer for its time, already it was less independent and less distinguished than the MESM. In subsequent years, the Soviet Union had more and more difficulty maintaining the pace of competition. Eventually, in the 1970's, it gave up the attempt to develop an independent series of manufactured computers and shifted over to IBM architecture and standards. After that it fell further and further behind.

The history of computers in the Soviet Union is a graphic example of the failure of that country to blaze an independent path, another case of the erosion of Soviet exceptionalism. The Soviet Union was forced to adopt Western patterns and to permit the breakdown of its controls over information. To have banned computers entirely, the only alternative, would have condemned the country to an even more hopeless obsolescence. Something similar happened in their space program, where early brilliance was followed by obsolescence, increasing adoption of Western designs, and, eventually, cooperation.

Gradually, Soviet leaders and economists became aware that the main reason for their failure to compete with the West technologically was not because their scientists and engineers were technically inferior. It was because Western economies were engines of change, dramatically driving the pace of innovation. Once Soviet leaders realized that they had to make economic changes in order to compete technologically, the claim for Soviet exceptionalism became less and less credible. The creation of joint ventures with Western firms in the last years of the Soviet period was a recognition that the only hope for survival of the Soviet Union was in cooperation with the West, not in defiance of it.

In *domestic politics*, the rise of environmental worries in the Soviet Union caused by the damage of industrial technology was a powerful force for a convergence of Soviet and Western societies. In the history of the decline of the Soviet Union, the great public debate over the fate of Lake Baikal that began in the early 1960's has a significant place. As Paul Josephson and Douglas Weiner have shown in recent studies, scientists in the Irkutsk region began public protests against proposed incursions on Baikal as early as 1958, just five years after the death of Stalin.[8] In this discussion, it became obvious that the environmental degradation that Soviet ideologists had earlier linked uniquely to rapacious capitalism was also endemic to Soviet socialism. At the same time, embryonic forms of Western-style interest groups, whistle-blowers, and investigative reporting began to emerge. In 1965, workers at the site of the Baikalsk rayon plant wrote a letter to *Literaturnaia gazeta* complaining that waste from the plant was being dumped directly into the lake, bypassing the settling tanks.[9] In the United States, such action by workers over the heads of their supervisors would be called "whistle-blowing." A correspondent for *Literaturnaia gazeta* named Oleg Volkov made the Baikal issue his own and went to the Baikalsk plant to find out for himself what was going on.[10] In the West, we would call this "investigative reporting."

In the debate that followed, different parts of the Soviet bureaucracy took different positions. The ministry in charge of the pulp and wood-processing industry defended the plant managers, but the ministry in charge of fisheries criticized them for endangering the fish harvest of Lake Baikal.[11] Environmentalists, sportsmen, scientists, and nature lovers began to establish embryonic forms of what in the United States we would call interest groups. And the spokespeople for these groups thoroughly scrambled traditional ideological lines; the conservative journal *Oktiabr'* opposed the pollution of Lake Baikal, but so did the traditionally liberal *Novyi Mir* and *Literaturnaia gazeta.*

Critics wanted to know why officials who were in charge of increasing production at the plant were also responsible for environmental protection.[12] The critics saw a conflict of interest here, and they called for the creation of a separate ministry of the environment with regulatory powers. Here we see a recognition that the all-powerful Communist Party could not be relied upon for the solution to all society's problems, and we also see an incipient recognition of the need for separation of powers and functions, embryonic pluralism.

Behind all the details of these discussions lurked a fundamental issue that has not been properly noticed. A technical issue—pollution by industrial effluents—was providing cover for political, national, and emotional questions that would not otherwise have been able to get into the official press. Perhaps the most exaggerated example of this hiding behind technical issues to obtain political goals came when a journalist maintained in *Pravda* that the water of Lake Baikal was so pure that "it may be needed within the very near future for research in the physical properties of matter."[13] No one pointed out that water cannot be purer than the distilled water produced in laboratories; the usefulness of the scientific falsehood was that it allowed one to cloak a political question in the garb of objective science and therefore legitimately criticize the regime.

When independent-minded people in the Soviet Union noticed that one could criticize the Soviet government and the Communist Party in the name of environmentalism and get away with it, they flocked to the environmentalist cause. Environmentalism thus served as a surrogate for political dissidence. In the Baltic republics, for example, a great impetus to the nationalist movements was given by environmentalists who criticized the damage to nature in the Baltic region caused by Moscow industrial bureaucrats.[14] Once *perestroika* and *glasnost'* arrived, and it became possible to speak frankly on issues other than the environment, the strength of the environmentalist movement waned. Protective cover was no longer needed, but it had played a very significant role in giving birth to independent political movements.

In each of the areas discussed above—scholarship, information media, foreign policy, economic development and domestic politics—science and technology acted as moderating or leveling forces reducing the distinctive characteristics of Soviet society. It is now appropriate to ask, "Do science and technology *always* act in this way? Is there something inherently westernizing about science and technology? Was it inevitable that science and technology would draw Russia more and more into the Western orbit?" A simple affirmative answer to these questions is not, in my opinion, appropriate, nor would it be true to my earlier comment that I support an "interactionist" model of the relationship between science and society, not a determinist one. I do not believe that the moderating and westernizing effects of science and technology on the Soviet Union were inherent or inevitable, but rather that they were a result of the particular circumstances of Soviet society and the place that science and technology occupied in it.

During the Soviet period, science and technology were leveling influences that brought Russian society and Western society closer together. In the post-Soviet period, the scientific and technical intelligentsia in Russia is fracturing, and some members of it are now advocating positions that are, intentionally or unintentionally, resulting in a strengthening of anti-Western positions.

A major event in the discussions among scientists in post-Soviet Russia about their attitudes toward the West was an interview, entitled "I Will Never Return to Russia," on May 5, 1993, by a correspondent of *Izvestiia* with Academician Aleksei Abrikosov, a prominent physicist who had emigrated permanently to the United States. The interview caused a sensation and was later published in a condensed form on the op-ed page of the *New York Times*.[15]

In the interview, the *Izvestiia* reporter noted that many outstanding scientists such as Abrikosov had emigrated to the West, causing a great brain drain, and asked "Do our specialists dream of returning home?" Abrikosov answered: "How can you seriously ask about Russian scientists' nostalgia for their homeland when in Russia nobody needs science? Fundamental science does not bring a quick profit, it requires budget financing, but there is no money for science in the state budget."[16] The correspondent then asked Abrikosov if he would encourage Western foundations and Western scientists to lend a helping hand to Russian scientists in this moment of crisis.

Abrikosov replied:

> I am certain that helping science in Russia is hopeless. . . . Today there is only one way to preserve Russian science: help all talented scientists to leave Russia as soon as possible, and forget about the rest. You think that I am exaggerating? Many people argue with me. But life shows that I am right. Not one of the joint international projects in science has worked out. On the contrary, any Russian scientist will leave the country as soon as he has the possibility.[17]

The publication of this interview, both in Russia and the United States, evoked a storm of criticism of Abrikosov. Academician Aleksandr Migdal, currently holding a tenured position at Princeton University, castigated it as "irresponsible" and "hypocritical" and said that Abrikosov's suggestions "smell like Stalinism."[18] Yakov Alpert at the Smithsonian Observatory at Harvard called Abrikosov's words "immoral" and "cynical," and expressed the fear that American readers would be disoriented by them and would cease trying to help Russian science.[19] The editors of *Izvestiia* could not let Abrikosov's words be published uncriticized. They attached a commentary calling them "amoral" and asked rhetorically, "Where should yesterday's Soviet scientists live and work—in their own country or abroad? For authentic Russian scientists, devoted to science in the country that nourished them, there cannot exist two answers to this question."[20] It was an editorial comment that reminded one of official Soviet condemnations of dissidents and refuseniks in the Brezhnev period.

The Abrikosov interview obviously touched deeply a nerve of national pride among many readers. It provided ammunition to nationalist critics who wished to portray scientists as unpatriotic and disloyal.

A somewhat similar issue concerning science has been the controversy in Russia over the millions of dollars of aid to science provided by the Western financier George Soros. Although Russian scientists have by and large been very grateful for this assistance, Soros became a controversial figure and a target for nationalist critics. In February 1994, the right-wing Russian journal *Molodaia gvardiia* published an article entitled "The Secret Diplomacy of George Soros."[21] The author of the article, S. Moskvitianin, accused Soros of trying to set up a secret world government, working together with David Rockefeller, Henry Kissinger, and Mikhail Gorbachev. Moskvitianin concluded his article, "Only the patriotic forces and a national government can put an end to this ruinous duplicity, stop the secret diplomacy, and free the country from the bondage yoke of the transworld government."[22]

It was one thing for Soros to be criticized in such publications as *Molodaia gvardiia*, well known for its crude anti-Semitic and nationalistic stances. Most Russian readers dismissed such diatribes. But on January 10, 1995, the respected newspaper *Nezavisimaia gazeta* published an article that was purportedly an internal document of the Russian Federal Counterintelligence Service (FSK) accusing American foundations, universities, and volunteer organizations of carrying out espionage and subversion.[23] Among the organizations accused of espionage were the American Association for the Advancement of Slavic Studies, the Peace Corps, the American Council of Learned Societies, the International Research and Exchanges Board, the Ford Foundation, and Harvard, Yale, Columbia, Duke, and Stanford universities. But George Soros and his International Science Foundation were given special attention. The report accused the ISF of espionage, of promoting a brain drain, and of an effort to "doom Russian science and the Russian economy to lag behind, to be controlled by America."[24]

In the months since this sensational article appeared, there have been dozens, if not hundreds, of articles and letters to the editor published on the subject in Russia.[25] Most of them have supported the Western philanthropies. The effect of the controversy has been a rallying of support by the Russian intelligentsia around George Soros and the International Science Foundation. But not all suspicions of Soros and the ISF have disappeared. On March 13, 1995, Aleksandr Solzhenitsyn said in an interview on Russian television that Russia is letting itself become an "ideological colony" of the West. He said that Russia needs to boost its "ideological defense against scientific and cultural grants from the Soros Foundation."[26] The controversy has not ended.

Meanwhile, the scientific-technical intelligentsia in Russia has been undergoing a political evolution as its economic position changes. At the time of the breakup of the Soviet Union, the majority of scientists and engineers in that country favored Western-style reforms. They called for the de-ideologization of science, its demilitarization, and its democratization. Some progress was made on each of these fronts. And what has been the result? For most Russian scientists and engineers, the result has been impoverishment and unemployment. The plight of research scientists is often desperate; salaries have not kept up with inflation, and, consequently, most researchers—if they get paid at all—make less than taxi drivers and small traders, who are in a better position to keep up with the market. Scientists and engineers in Russia are losing their illusions about the benefits of democracy and a free market. Only a few years ago, they were among the most ardently pro-Western of all citizens in the former Soviet Union, and they would back any party that called for democratic reforms and a free market. Now they are increasingly ready to back any party that will support science, regardless of the other ideological commitments that party may have. On November 4, 1995, the leading newspaper of the R&D community, *Poisk,* carried an article on its front page entitled "Where Are You, the Party of Science?" which called for scientists and engineers to support any party that would come to the aid of science, and specifically called for help from the Communist Party of the Russian Federation and the Congress of Russian Communes, both antireform parties.[27] In response, leaders of nationalistic and communist parties, including Vladimir Zhirinovskii and G. A. Ziuganov, have recently called for additional government support for science and technology, especially military technology.

The political leader who has been most active in addressing the needs and frustrations of the technical intelligentsia is Ziuganov, the head of the Communist Party of the Russian Federation. An educated man himself, with a background in mathematics and philosophy, he has made a direct appeal to intellectuals. He has adopted as the official seal of the new Communist Party, in addition to the traditional hammer and sickle, the "book," adding scientists to workers and peasants as bulwarks of the political order he seeks to create. In his speeches and writings, Ziuganov has called for the preservation of the necessary technical foundation for the military defense of Russia's interests, an appeal that is certain to please the engineers and technical personnel in Russia's enormous but now grossly underutilized military industrial complex. Ziuganov has combined this interest in engineers and scientists with reliance on patriotic and Slavophilic doctrines from the past, and has approvingly cited Nikolai Danilevskii and Konstantin Leont'ev, calling for the revival of "the Russian Idea." Thus, Ziuganov is creating an ideological program that is designed to split the Russian intelligentsia, bringing the more conservative, nationalistic, and provincial elements—particularly strong among the engineers and the technical intelligentsia—over to his side, while gladly abandoning the more reformist, liberal, and westernized intellectuals to Grigory Yavlinskii and the "Iabloko" party. And in making this effort, Ziuganov has allies among some highly placed scientists. When, in early June 1996, Ziuganov made a campaign stop at the science city Akademgorodok in Novosibirsk, the head of the Siberian branch of the Russian Academy of Sciences, Academician V. A. Koptiug, arranged for his speech to be carried on loudspeakers in the main street, a favor not accorded any other candidate. Shortly after his visit, a university professor in Novosibirsk reported that her faculty was splitting into factions, "the democrats and the communists."[28]

The faction of the Russian scientific-technical intelligentsia most susceptible to anti-Western political solicitations are the scientists and engineers from the military-industrial complex, the professional people experiencing the deepest cuts in recent years. A dramatic indicator of their distress was the suicide, on November 1, 1996, of Vladimir Nechai, the director of Russia's Cheliabinsk-70 nuclear weapons center. In his suicide note, Nechai stated that he could no longer watch his life's work fall apart, and that he was ashamed to face the people at his center, who had not been paid in five months.[29]

Even before the demise of the Soviet Union, engineers in the military sector were the most conservative portion of the scientific and technical intelligentsia. They were intimately involved with the Soviet system and the military-industrial complex, and they were often among the most stalwart supporters of the Soviet economic and political order. These engineers were by far the larger part of the Soviet technical intelligentsia.

The Soviet Union had more engineers than any other country in the world—almost 800,000 specialists, about 80 percent of whom worked for the military-industrial complex.

Soviet engineers were participants in "The Big Deal" of which Vera Dunham has written so eloquently, an arrangement reached during and after World War II that guaranteed employment and a range of limited but much-desired perquisites for Soviet technical specialists in return for their loyalty to the regime.[30] And the engineers largely lived up to their part of the bargain. The dissidents of the 1960's and early 1970's, writing critiques of the regime for publication in *samizdat*, were very rarely engineers; instead, they were typically natural scientists or members of the literary intelligentsia. The engineers were remarkably passive politically, compared to other professionals. No wonder engineering was sometimes called "the gray profession."

Many of the engineers from the defense industries took pride in the status of the Soviet Union as a superpower and in their contributions to that international eminence. If they had private complaints about the Soviet system, they centered on the relatively low social status of engineers and the bureaucratic barriers they often encountered when they tried to introduce technical innovations. When Gorbachev came to power in 1985, at first he excited the engineers by calling for technical modernization and economic acceleration (*uskorenie*), as well as higher salaries for engineers. But the hopes of the engineers for Gorbachev and for reforms were soon dashed. The promise of emphasis on technology and engineers went unfulfilled, and the economy of the Soviet Union continued to decline. Then, at the end of 1991, the Soviet government died, and the USSR no longer existed. The secure world of the engineers in the military-industrial complex crashed to the ground.

According to one Russian study,[31] the number of researchers employed in military institutes dropped 40 percent between 1985 and 1992, and another 20 percent drop is estimated to have occurred since. Unemployment is higher among engineers from the defense industries than any other profession.

In response to their desperate situation, the engineers have, at last, begun to become more active politically. A group of engineers in St. Petersburg has begun circulating a manifesto calling for the formation of a new ruling class made up primarily of engineers, whom they describe as "the most appropriate professional group in their social characteristics, the most numerous, and who possess the most creative potential." In their manifesto they take an anti-Western position:

> Russia is not the West, and not the East. It is distinctive, and the loss of its distinctiveness would mean its ruin. Attempts to westernize Russia have been made more than once, and all of them have brought more harm than good. The present reforms, the most radical and the most unsuccessful, are a blind attempt to westernize Russia. The attempt can lead to its ruin.[32]

Conclusions

The shifts in attitudes of Russian scientists and engineers toward the West illustrate the weakness of a deterministic interpretation of the social roles of science and technology and the strength of an "interactionist" model. At different times and in different social contexts science and technology have very different effects. When the totalitarian and xenophobic features of the Stalinist Soviet Union were strongest, science and technology, with their emphasis on objective knowledge and international connections, tended to have liberating and westernizing effects. However, with the collapse of the Soviet

Union and the disappearance of strong state support for scientific institutions, Russian scientists and engineers find themselves in a fight for survival. For some of them, it is tempting to blame their present predicament on the advent of a market economy. The science that existed under state socialism looks more and more attractive. The scientific and technical community in Russia is fracturing under these pressures, with parts of it remaining loyal to their old democratic and Western affinities, despite their current difficulties, and other parts drifting toward communist, nationalist, and anti-Western positions.

In conclusion, I would observe that while scientists and engineers played the role in the Soviet period of reducing the ideological exceptionalism of that country and bringing it closer to the West, in the current situation, the roles of scientists and engineers are much more complex. Those scientists who have emigrated to the West or who favor cooperation with Western universities and companies are increasingly becoming targets of the nationalists and communists, who are simultaneously beginning to court those scientists and engineers who remain behind, promising to support them in a way in which the free-market reformers of the Yeltsin government are not. Thus, it is not inconceivable that in the near future, an alliance will be created between the opponents of Western-style reforms and an important portion of the scientific and engineering community. If that happens the result will be a loss both for Russia and the West.

CHAPTER THREE

How Robust Is Science Under Stress?

Scientists studying nature in the laboratory frequently subject their objects of research to extreme conditions. By studying matter at very low or very high temperatures or pressures, or by accelerating bits of matter to incredibly high velocities, or by killing and dissecting previously living fauna or flora, or by testing animals under conditions of stress, scientists hope to draw conclusions that will be informative about nature in general. The study of the extreme is often thought to enlighten our understanding of the normal.

Historians and sociologists of science are in a very different position. The object of their study is science itself, but they obviously are not permitted to submit scientists and the research process to artificially contrived and extreme conditions in order to see how the scientists will react and how the research process will be altered. But if such extreme conditions could be found in recent history, they might nonetheless tell us something about science.

Very few people outside of the former Soviet Union fully comprehend the extreme conditions—of both favoritism and persecution—to which science there has been submitted during the preceding two generations. Even many citizens of Russia, including scientists themselves, are unaware of the full dimensions of the extraordinary experiment through which they and their parents have lived. The Soviet Union devoted a larger share of its budget to the support of science and technology than any other industrialized nation in the world, but it also persecuted its scientists in a way unparalleled in history. Since the fall of the Soviet Union, the scientists there have been given a large degree of political and intellectual freedom, but at the same time they have been subjected to a financial contraction that is equally unprecedented in history. How robust is science under severe stress? Which is more important to science, money or freedom? The Soviet and post-Soviet periods provide sobering, even appalling, test cases for examining these questions.

For those people who believe that science is a fragile flower, an intellectual endeavor that can easily be destroyed, the Soviet experience provides a daunting challenge. During years in which the Soviet Union was building the world's largest scientific establishment, conducting research that would garner Nobel prizes, building missiles and nuclear weapons, it was also persecuting its scientists and engineers mercilessly. To give a list of all Soviet scientists who were repressed by the secret police would be not only impossible but tedious. I list some of the most prominent below. Soviet scientists who were arrested by the police and accused of grave crimes included people, who, before or after their arrests, were the designer of the Soviet Union's most famous airplanes;[1] the main theoretician of the Soviet hydrogen bomb;[2] the head of the Soviet space program (who directed the launching of the world's first artificial satellites);[3] three Nobel Prize-winning physicists;[4] several of the giants of the development of population genetics;[5] three successive presidents of the Agricultural Academy;[6] the director of world-famous Pulkovo Observatory;[7] the director of the Institute of Experimental Medicine;[8] the director of the Institute of Mathematics and Mechanics of Moscow University who was also the president of the Moscow Mathematical Society;[9] the director of the Azerbaidzhan Institute of Microbiology;[10] the director of the Leningrad Astronomical Institute;[11] the head of the department of plant genetics at Leningrad University;[12] the director of the Institute of Soil Science;[13] the chair of the faculty of medicine of the First Moscow Medical Institute;[14] a vice president of the Ukrainian Academy of Sciences;[15] the director of the Institute of Physics at Moscow University, who was also a pioneer in the development of "externalism" in the history of science;[16] the director of the Tashkent Observatory;[17] the director of the Siberian Institute of Grain Culture;[18] the director of the Mechnikov Institute of Infectious Diseases;[19] the director of the Institute of Agricultural Soil Science of Belorussia;[20] two different directors of the Khar'kov Physico-Technical Institute;[21] a forerunner of animal and plant ecology;[22] two different rectors of the Moscow Higher Technical School (better known as the "Bauman Institute," one of the leading engineering schools of Russia);[23] the academic secretary of the Academy of Medical Sciences;[24] the first woman ever elected to full membership in the Academy of Sciences of the USSR, an outstanding physiologist;[25] the director of the All-Union Institute of Stock Breeding;[26] the director of the Vladivostok branch of the Institute of Chemistry;[27] the director of the Ukrainian Institute of Applied Botany;[28] the director of the Leningrad Blood Transfusion Research Institute;[29] the head of the trade union organization for engineers and technicians;[30] the director of the State Microbiological Institute;[31] the chairman of the Russian Technical Society;[32] a vice president of the Belorussian Academy of Sciences;[33] the director of the Institute of Medical Genetics;[34] the director of the Institute of the History of Science;[35]

the dean of biology of Moscow University;[36] the director of the Institute of Rare Metals;[37] the director of the Institute of Hybridization;[38] the director of the Geological Institute;[39] the dean of chemistry of Moscow University, a leader in chemical catalysis;[40] the director of the Central Laboratory of Ionization;[41] the chairman of the Association of Microbiologists;[42] and many, many more. Many of these scientists either were shot or died in labor camps.

And these are only a few of the most famous. The persecution ran through the ranks of average researchers and designers as well. No field of knowledge was spared. Probably half of the engineers in the Soviet Union in the late 1920s were eventually arrested. The percentage of scientists arrested may never be known, but it was probably lower than among the engineers. We do know that in autumn of 1928, during just a few weeks, 648 members of the staff of the Academy of Sciences were purged. According to the official figures published by the secret police,[43] 19 percent of the staff personnel in the departments surveyed were fired or seized at this time. And the peak of the purges did not come until 1937! Approximately 20 percent of all Soviet astronomers were arrested in 1936 and 1937.[44] Arrests of scientists continued until Stalin's death in 1953. Until 1956 or 1957, some of the Soviet Union's most eminent scientists and engineers worked in prison laboratories.

Even now, decades after these events, anyone who looks over this record will be appalled. When Charles Gillispie, a historian of science at Princeton University, reviewed a book of mine in which I described this unimaginable persecution, his reaction was, understandably, to wonder "how science could function at all."[45] The question is an important one that raises disturbing implications for those people who think that freedom is necessary for science to prosper. For me this difficult question was raised some years ago in a very poignant way.

In the 1970's, while I was doing research on the history of Soviet science that was revealing this record, I was asked to serve on a panel of the National Academy of Sciences in Washington, D.C., that was evaluating the quality of Soviet science. Most of the members of the panel were distinguished American natural scientists who were familiar with Soviet work in their fields.[46] I was invited to participate in the study as a historian of Russian science (and also as rapporteur for the report). I was the junior member of the panel and, as an engineer-turned-historian rather than a natural scientist, was considered by the other members of the panel to be something of an outsider. My job was to help conduct the two-year study, which included sending questionnaires to several hundred American scientists who had worked with Soviet colleagues, to listen to the reports of the natural scientists, record them, and write summary evaluations of Soviet science for the other members to either approve or disapprove. I took the job seriously, and even today I have the papers and reports.

The striking aspect of the evaluations given by American scientists of Soviet work in their fields was the excellence they found there. Lipman Bers, past president of the American Mathematical Society, professor of mathematics at Columbia University, and a native reader of the Russian language, described Soviet mathematics in the 1970's as equal to mathematics in any other country in the world, and said that Moscow probably contained more great mathematicians than any other city anywhere. Bers was fully aware of past and continuing persecution of scientists in the Soviet Union (this was the time of refuseniks in Soviet mathematics, the firing of mathematicians who applied to emigrate to Israel). But, he said, the strength of Russian mathematics was so great and the pool of talent so deep that repression could not destroy it.

In a similar mood, David Pines, professor of physics at the University of Illinois, reported that "in the area of theoretical condensed matter physics, USSR scientists have been doing some of the most innovative and important work in the world for two decades or more." Hans Frauenfelder, another physicist from Illinois, wrote that "the level of the best work in solid state theory in the USSR is outstanding and at or near the frontier."

Similarly strong comments were made about other fields in the field-by-field reports submitted by outstanding American scientists, most of them members of the National Academy of Sciences:

On solid-state chemistry: "This is a field in which the Soviets have it and the United States simply does not."
On arc welding: "The USSR is at the world forefront."
On electronic materials: "Soviet work . . . is at a level comparable to if not more advanced than ours."
On plasma physics: "Soviet scientists are at the very forefront in device construction and in pure theory."
On theoretical seismology: "The USSR has been a world leader for half a century."
On climate research: "The USSR is a world leader."
On theoretical astrophysics: "The Soviets are universally regarded as being at the very forefront of world efforts."

In order to establish how successful exchange trips by American scientists to the Soviet Union had been, questionnaires were sent out to all American scientists who had participated in these exchanges since their beginning in 1959, a total of about 350 American scientists. It is worthwhile noticing that these exchanges began only six years after the death of Stalin, and only three years after Khrushchev had released many Soviet scientists from prisons. (I went to the Soviet Union myself for an academic year starting in 1960, four years after the prison release, and I talked to a number of Russian scientists who had recently returned from the camps.) Of the 350 American scientists sent questionnaires, 275 (78.6%) answered. Evaluating their trips to the Soviet Union, 32.4 percent called the results "outstanding"; 42.3 percent, "very good"; 18.8 percent, "satisfactory"; 4.8 percent, "fair"; and only 1.8 percent, "poor."

Despite the fact that 75 percent of all respondents rated their experience in the Soviet Union in the late 1950's, 1960's, and early 1970's as "outstanding" or "very good," it should not be assumed that these American scientists were uncritical or naive about the political realities of the Soviet Union. Having lived in the Soviet Union for many months, most of them knew about political restrictions there. When asked to agree or disagree with the statement "The scientific productivity of the exchange is hampered by the political situation in the Soviet Union," 25.4 percent responded "strongly agree"; 47.3 percent, "agree"; 24.2 percent, "disagree"; and only 3.1 percent, "strongly disagree."

Furthermore, almost all of the American scientists agreed that Soviet science was not as distinguished as American science, and many of them believed that political restrictions were important reasons why the enormous Soviet science establishment failed to fulfill its potential.

Taking all these results together, one would have to conclude that American scientists in close contact with their Soviet colleagues were, on the whole, quite aware that politics was interfering with Soviet science, but they nonetheless thought that this interference was not so severe that it prevented valuable scientific work, including collaborative work, from being done. When one considers that the Soviet science establishment in the previous 30 years had passed through the most repressive experiences in the history of modern science, the fact that so much good science had survived is truly amazing and demands further explanation and analysis.[47] But before I try to give such an explanation and analysis, I would like to move forward to the next great crisis in Russian science, the one that came when the Soviet Union collapsed, a little more than a decade after the publication of the National Academy of Sciences study I have just described.

When the Soviet Union disappeared in December 1991, the system of financial support for science in that country disappeared with it. Total government expenditures on "science and scientific services" dropped from 1.03 percent of gross national product in 1991 to 0.52 percent in 1993 to 0.3 percent in 1996.[48] At the same time, the size of the gross national product was also declining precipitously. Between 1992 and 1996, the financing of the Russian Academy of Sciences, the home of fundamental science, decreased by more than ten times. The salaries of scientists in the Academy decreased, in real terms, more than five times. Most institutes in the Academy almost stopped buying new equipment and sponsoring expeditions in order to save what little money they had for the salaries of researchers. In 1995, the highest salary paid by the government to researchers in the fundamental sciences—that to a full member of the Russian Academy of Sciences—was 450,000 rubles a month, about U.S.$94 at the contemporaneous exchange rate. Average researchers received much less. In 1995, the salary for a "senior research worker" was established at 75,000 rubles a month, about $15. The monthly stipend of a graduate student at the Moscow Physico-Technical Institute, an elite institution, was 50,000 rubles a month, about $10, in 1995. Even worse, many of these stipends, meager as they were, were simply unpaid for months at a time. In 1994 and 1995, approximately 80 percent of the researchers in the Russian Academy of Sciences went "on leave," receiving little or no salary.[49] The best of these researchers found support from foreign foundations.

Foreign foundations gave important, but probably temporary, support in the 1990's. The most important of these foundations was the International Science Foundation (ISF), established by the New York financier George Soros specifically to support science in the former Soviet Union. In one of the most remarkable chapters in the history of philanthropy, Soros supplied over $130 million in the years 1992–95 for the support of science in these countries.[50] (I was a member of the Executive Board of the ISF and may therefore not be the most objective analyst of its activities.) In 1993, Soros's International Science Foundation gave more money to the Russian Academy of Sciences than the Russian government itself,[51] and in 1995, after the input from Soros to science began to diminish, foreign foundations still supplied about a third of the support of Russian science.[52] Foundations other than the ISF that were important included the International Association for the Promotion of Cooperation with the Scientists from the New Independent States of the former Soviet Union (INTAS), supported by the European Community; Soros's Open Society; the John D. and Catherine T. MacArthur Foundation; the Howard Hughes Foundation; the International Science and Technology Center, funded by European states, the United States, and Japan; and the Civilian Research and Development Foundation, funded by a consortium of foreign countries and private foundations.[53] However, in 1994–95, Soros curtailed his support for science, the ISF ceased most of its activities, and it became ever more apparent that science in the former Soviet Union would have to survive on its own resources. By the first half of 1995, the role of foreign foundations in funding Russian R&D had diminished. By 1996, most of these expenses were again being met by the Russian government, and a very small amount was provided—less than a tenth of what had been available before the collapse of the country.[54]

Many researchers, benefiting from the liberal travel regulations that came with the disappearance of Communist Party control, fled abroad, either permanently or temporarily. The exodus was greatest among the very best scientists in the former Soviet Union, because they had much better opportunities than their less well known colleagues to find positions abroad—visiting professorships, fellowships, and tenured professorships. The scientists and engineers who remained in the former Soviet Union, especially the young, often sought economic refuge by deserting the laboratories for private business activity, where their scientific talents were not usually utilized. This "internal brain drain" was much larger than "external brain drain," since not many scientists in the former Soviet Union received offers from foreign institutions, but quite a few were able to find private economic activity at home.

Journals and newspapers in Russia were filled with articles describing the "destruction of Russian science" and even predicting "the death of science."[55] While these articles contained exaggerations and were written in the apocalyptic style so common in the post-Soviet press, no one knowledgeable about the situation, either in the West or in Russia, denied the existence of a deep crisis. In fact, *crisis* may be too weak a word to describe the situation in Russian science in the 1990's. As V. E. Zakharov and V. E. Fortov, two prominent members of the

Russian Academy of Sciences, wrote in an article at the end of 1994:

> Today it is generally recognized that our science is experiencing a crisis. However, this is far too optimistic an assessment of the problem. It is our deep conviction that the danger currently threatening our science is unprecedented. Our science was not in such danger during the Civil War, nor during World War II, nor at any time during our history. In just three years, shock therapy has dealt our science such a series of blows that it would be more correct to say that our science is not in crisis, but in a comatose state.[56]

It is striking that Zakharov and Fortov, writing at a time when political restrictions had disappeared, did not even mention the Stalinist repressions of Soviet science as a great crisis, referring instead to problems brought by the military emergencies of the Civil War and World War II.

Obtaining good statistics on just how many scientists from the former Soviet Union have emigrated is extremely difficult. No agency collected reliable and complete statistics. Even when we do have statistics, we do not know exactly what to make of them, since scientists who leave the former Soviet Union do not usually say whether they are leaving permanently or temporarily, and often they themselves do not know.

According to the U.S. State Department, between 1990 and 1993, about 10,000 scientists and engineers emigrated from the former Soviet Union to the United States. In 1990, there were approximately 2,000; in 1991, 2,500; in 1992, 3,000; and in 1993, about 2,000.[57] The OECD in Paris estimated that by the autumn of 1993, just under 30,000 Russian scientists had emigrated to all countries. This much larger figure than that supplied by the State Department includes all emigrants with undergraduate or graduate degrees in the sciences and engineering; the number of them who were, at the time of emigration, active researchers with graduate degrees in the natural sciences was considerably smaller. Glenn Schweitzer, of the staff of the National Academy of Sciences in Washington, D.C., estimated that by 1995, only about 2,000 active researchers in the natural sciences with advanced degrees had left, a very small percentage of the R&D workforce, although probably including some of the best.[58] Internal emigration, the abandoning of scientific careers in the former Soviet Union, has been much more significant. For every Russian scientist who emigrated permanently abroad, ten left science for another activity inside the country.[59] According to the OECD in Paris, between 1989 and late 1993, the number of scientific and technical workers in the former Soviet Union declined by over 50 percent, from 1.5 million to just over 600,000. The decrease was greatest in industrial and military areas, much less in fundamental science. For example, the Academy of Sciences of the USSR, the traditional seat of fundamental research, had 160,000 researchers in 1990; its successor, the Russian Academy of Sciences, had 130,000 in 1995, a decline of not quite 20 percent.

If we look at individual institutes, we find a very uneven picture of emigration; some laboratories and institutes were hit very hard, others not much at all. The differences seem to be explained by opportunities abroad and institutional cohesion at home. By early 1995, approximately 20 percent of the senior researchers of the Moscow Institute of Molecular Genetics had left. In the Lebedev Physics Institute's theoretical division, of 55 scientists, 5 had emigrated permanently by mid 1995 and another 10 were abroad, their return uncertain. In the Kurchatov Institute of Atomic Energy, almost 40 percent of the plasma-theory scientists had left for the West by mid 1995. In the Institute of Cytology in St. Petersburg, 50 of the institute's 250 scientists were abroad by mid 1995. On the other hand, at the Ioffe Physico-Technical Institute in St. Petersburg, only 60 out of 1,200 were abroad in March 1995. At the Komarov Institute of Botany in St. Petersburg, very few senior researchers had left by early 1995.

The brain drain hit Russia's space industry especially hard. The Russian Space Agency's general director, Iurii Koptev, estimated that the Russian space industry had lost 30 percent of its highly trained specialists by mid 1994. When the Russian space mission to Mars failed in 1996, the scientists in charge announced that there was no money available to try again.

Scientific publication declined precipitously. The output of the publishing house "Nauka," the press of the Russian Academy of Sciences, dropped by 1995 to about one-fifth of its previous level. By the mid 1990's, it was not unusual to find scholarly journals in the former Soviet Union that instead of publishing the usual six or twelve issues a year published only one or two, perhaps labeling each copy "Issues 1–6," indicating that six issues were combined in one. Subscriptions to foreign journals were drastically curtailed.

How did the research institutes and laboratories in Russia respond to such dramatically decreased budgets? In 1995 and 1996, I visited several leading research centers in Moscow and St. Petersburg: the chemistry faculty of Moscow University, the Ioffe Physico-Technical Institute in St. Petersburg, and the Komarov Botanical Institute in St. Petersburg.[60] All three were trying to exist on budgets that were less than one-tenth of what they had had five or six years earlier. Yet they seemed to be surviving. How were they doing this? Let me take the example of the Ioffe Physico-Technical Institute. The director of this institute, Zhores Alferov, gave me some particularly revealing budgetary statistics.

The Ioffe Physico-Technical Institute is one of the most famous physics institutes in all of Russia and has often been called "the cradle of Soviet physics."[61] Three of its researchers have won Nobel Prizes: P. L. Kapitsa, L. D. Landau, and N. N. Semenov. The leader of the Soviet atomic project, I. V. Kurchatov, was also as a young scientist at the Ioffe Institute, as was Iulii Khariton, another prominent scientist in the atomic project. Although the record of the Ioffe Institute in recent years has not been quite as stellar as it used to be, it is still an outstanding center of physics research.

In 1990, the Ioffe Institute had a staff of 3,000, of whom 1,400 were scientists. By January 1995, the staff had dropped to 2,500, of whom 1,200 were scientists. In March 1995, approximately 60 of these 1,200 scientists were working abroad. Summing up the situation with personnel, the number of research scientists in the institute had dropped by only about 15 percent, or possibly 19 percent, if the scientists who were abroad did not return. Although this is a significant decrease, it is not nearly commensurate with the decrease in the overall budget of the institute, which had dropped much more dramatically.

Between 1990 and 1995, the budget of the institute decreased by about fifteen times. In rubles adjusted for inflation, in 1990, the budget of the institute was about 66 million rubles; in 1995, it was 3.4 million rubles. How could the institute survive in the face of such draconian financial cuts? The figure on page 66 shows how the budget of the institute changed in recent years.

Here we can see clearly what the administrators of the Ioffe Institute were doing in order to survive. In 1990, only about 7 percent of the institute's budget went for salaries; 45 percent of the budget went for materials and equipment. In 1995, almost

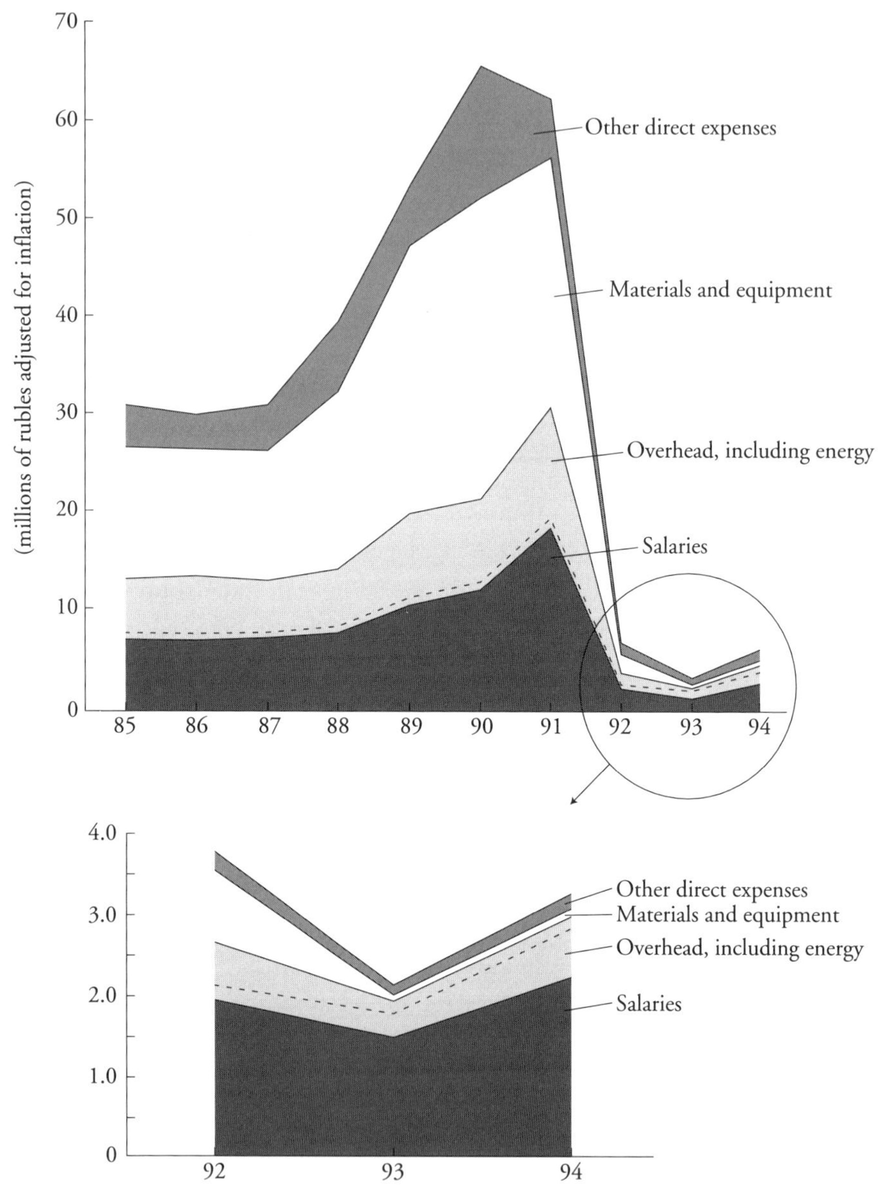

The Budget of the Ioffe Physico-Technical Institute from 1985 to 1994

70 percent of the institute's budget was going for salaries; only about 5 percent was going for materials and equipment. We can sum up the situation in the following way: the Ioffe Physico-Technical Institute was surviving by ceasing its purchases of materials and equipment and devoting its meager resources primarily to the payment of its staff. Such a policy is obviously a stopgap measure that could work only for a short while. For a few years, it is possible to stop buying instruments and materials, but if this pattern continues for any length of time, the laboratory will fall hopelessly behind. New equipment—computers and instruments of all kinds—are absolutely essential in modern scientific research.

By the middle and late 1990's, the pattern in the Ioffe Institute was being repeated all over the former Soviet Union. Research institutes were trying to hang on to their research staffs and in order to do so they were stopping almost all other expenditures. The overall staff of the Russian Academy of Sciences decreased only 19 percent during 1990–95, while the budget has decreased by almost 1,000 percent.

Conclusions

Science in Russia and the Soviet Union has passed through two great crises during the past 70 years. In the period 1929 to 1953, science there was subjected to political oppression of an intensity unprecedented in modern history, with tens of thousands of scientists and engineers arrested, imprisoned, and sometimes executed. In the years 1989 to 1997, it was subjected to a financial crisis, one still going on, equally unparalleled in modern history, with research budgets slashed to less than a tenth of what they had been. What do these traumatic experiences tell us about science?

The first conclusion that can be drawn from these dismaying experiences is that science is incredibly robust, able to withstand blows that almost anyone who heard about them beforehand, in the abstract, would assume would be mortal. Both during the Stalinist purges and during the post-Soviet economic contractions, Russian science suffered grievously, and its productivity and creativity were severely limited by these experiences. But the striking conclusion that must be drawn is that science can survive unbelievable punishment. Five Nobel prizes were awarded to Soviet physicists for work done in the 1930s and 1940s, a period of tyranny and terror. One of those physicists, Pyotr Kapitsa, had been kidnapped on Stalin's orders just three years before he did his most important research.

The second conclusion that must be drawn from these experiences may be deeply troubling to advocates of academic freedom. Which is more important to science, freedom or money? The Soviet and Russian experience would indicate that money is more important. All during the years of Stalin's terror, when the secret police were regularly arresting scientists and engineers, the Soviet government was pumping money into science and technology on a scale matched by few if any other governments in history, and Soviet ideologists emphasized the importance of careers in science and technology over all other professions. Student enrollments in the fields of science and engineering were growing at astronomical rates. For every scientist or engineer who was arrested, there were always several others ready to step into the empty shoes. Some of the best Soviet airplanes were designed in prison camps established for the purpose, and the Soviet atomic project, which resulted in the construction of successful nuclear weapons long before most Western leaders predicted, was run by the head of the secret police, Lavrentii Beria, one of the bloodiest policemen in history. Beria employed prison labor at the atomic weapons research establishments, such as Arzamas 16, where Andrei Sakharov worked for eighteen years; one rebellion by the inmates at Arzamas 16 (among manual laborers, not scientists) was suppressed violently, using army troops and mass artillery fire.[62] Every rebel was killed. Such ruthlessness must have done damage to science, but we must conclude that that damage was more than counterbalanced by the virtually unlimited financial and material resources Beria was given to build atomic weapons. Furthermore, he is credited by historians and scientists, both Russian and foreign, with being an effective, if cruel, manager.[63]

Sakharov later described in his memoirs how, in the years 1950–53, he watched through the window of his office at Arzamas 16 as columns of prisoners were marched by under armed guard.[64] He was disturbed by the brutality that surrounded him, but his concerns do not seem to have interfered with his scientific creativity. In these same years, Sakharov and Igor Tamm worked out an original toroidal approach to controlled thermonuclear reactions that still dominates the field today, in the United States as well as Russia and other countries.

Soviet scientists, despite the fact that many of their own family members and friends had been arrested or executed, faithfully served the system. We now know that the U.S. Manhattan project was plagued by a number of spies, scientists willing to turn over secrets to foreign powers.[65] There is no known case of a Soviet scientist in Beria's atomic project who betrayed the cause, even though the Soviet Union was a tyranny and its competitor the United States was a democracy. A strong moral case could be made for betrayal of the Soviet Union under Stalin, if there exist higher principles than political loyalty.

Why did Soviet scientists work so faithfully for a system that gave them such harsh treatment? This question raises deep and perplexing psychological issues, some of which will never be answered. There will always be something inexplicable about the Soviet scientists' commitment and loyalty in the face of such cruelty. Nonetheless, I think that some progress can be made in explaining this curious phenomenon. The first reason for their loyalty concerns the nature of the Soviet Union, and the second the nature of science.

Many Soviet scientists believed, or wanted to believe, that despite the appalling conditions in which they lived under Stalin, there was something important and salvageable in the Soviet experiment and in socialism. Certainly, in contrast to Hitler and fascism, they thought Soviet socialism had greater promise for human improvement if they could only live through Stalin's tyranny. In the memoirs of many Soviet scientists, one finds this underlying commitment to socialism and the Soviet effort to create a society different from that of capitalism.

The second reason for the scientists' loyalty to the Soviet Union has to do with science itself. If one is trying to live through a whirlwind of violence while retaining some sense of integrity, science (especially the more abstract areas in which Russians excelled) provides a refuge. In a perverse way, repression may actually have contributed to the energy and commitment that scientists gave to their profession. In the Soviet Union of Stalin's time, the world outside the laboratory or study was dangerous, and it also offered few attractive diversions. As a result, science almost totally absorbed many researchers' lives. Talented Soviet researchers sought to escape the politically threatening and morally corrupt atmosphere around them by submerging themselves in their work. Even if arrested, they sometimes continued their efforts in prison camps. Science was the one activity that made sense of their lives, the one area where they could serve truth without automatically coming into conflict with the system. Only physical annihilation could, and tragically often did, interrupt their devotion to science.

This commitment to science as a refuge from an unpleasant political environment may seem foreign to many American readers. However, in my own lifetime I have, at least at one moment, seen something a bit similar in the United States. In 1969–70, I spent a year as a fellow at the Institute for Advanced Study in Princeton, New Jersey, probably the most isolatedly cerebral institution in the United States. The year that I was there was perhaps the worst moment in the Vietnam War, a time when the country was racked by protests and antiwar demonstrations. During the year, U.S. soldiers massacred innocent Vietnamese civilians in the village of My Lai, causing a peak in the national outcries. Among the scholars at the Institute, many of whom were theoretical physicists or mathematicians, the news of these events caused considerable anguish, but a very common reaction was for them to bury themselves in their work in the hope that somehow all this would pass. The outside world was one in which they saw violence and untruth; in their work they could continue to pursue the noble ideals of scientific creativity and objectivity. The more reflective of these scientists were aware that their position could be questioned on moral grounds, just as were the more reflective Soviet scientists in Stalin's time, but the attractiveness of science under such conditions only grew.

The Soviet Union after Stalin came to possess almost half of the engineers in the world, and more than half of its geologists.

By the mid 1980's, the "density" of research scientists and engineers (full-time equivalent number of research workers per 10,000 inhabitants) in Russia was several times that in the United States.[66] If some diabolical social scientist had wanted to design an experiment to answer the question "Which is more important to science, freedom or money?" he or she could have hardly devised a more bizarre and telling comparison than that between the Soviet Union in its Stalinist years and Russia in its post-Soviet period. Surveying the damage done to Russian science in recent years, we must conclude that the financial crisis of the post-Soviet period has brought Russian science much closer to the edge of extinction than the political oppression of the Stalinist period. Russian science at the moment is barely hanging on, even though Russian scientists currently enjoy greater political freedoms than at any previous time in Russian history. In contrast to the Stalinist years, when science student enrollments burgeoned, student enrollments have decreased. Even those who graduate face formidable obstacles; of the graduates of Moscow University in 1995 in the sciences, more than three-quarters could not find jobs.[67] Many established scientists have either emigrated or deserted their laboratories for other more economically beneficial activities, desperately trying to survive. Equipment purchases, except those financed by foreign foundations or governments, have almost ceased.

Desperate as the current situation of Russian science is, however, I do not agree with those people who have announced its death. Russia still contains many young students and established scientists who are dedicated to research and teaching, and who will continue their work under almost any conditions. The robustness of Russian science is impressive today, just as it was in the time of Stalin's oppression. Still, one must admit that the economic crisis of the 1990's has come closer to stopping Russian science than the political repression of Stalin's time, an admission that may force us to rethink our assumptions about the conditions necessary for science's development.

The Russian example shows that, in extreme cases at least, money is more important for science than freedom, but of course the best science will be done when both freedom and money are present. American scientists are so accustomed to both freedom and money that they have lost their ability to compare the importance of the two. There have been moments when American scientists have experienced economic deprivation or political repression—they met economic deprivation during the Great Depression of the 1930's and during the decline of federal science budgets after the end of the Cold War, and some of them suffered political repression during the political witch-hunts of Senator Joseph McCarthy during the Cold War itself—but these experiences were so far from the Russian ones in their intensity that they can hardly be used for comparison. The Russian experience is truly unprecedented, and for that reason it should attract attention.

CHAPTER FOUR

How Willing Are Scientists to Reform Their Own Institutions?

From the late 1980's through the early 1990's, a major effort, or series of efforts, was made to reform the organization of science in Russia. The old system had been inherited from the Soviet Union and still displayed many characteristics of Stalinism. At a time when other institutions throughout the former Soviet Union were being swept by change, it seemed reasonable and logical that fundamental science, and especially the Academy of Sciences, would also be reorganized. Today, we can see that this effort failed. The Russian Academy of Sciences is at the present day largely a carbon copy of the old Academy of Sciences of the USSR, distinguished from its predecessor primarily by poverty, not by organizational reform.

How did this happen? Why did fundamental science prove to be one of the most conservative institutions of the old Soviet Union, able to resist the waves of criticism that swept over it during the last months of the Soviet regime and the first months of the new Russian one? In this chapter I shall try to answer this question by analyzing the history of the relations between the Soviet and Russian governments, on the one hand, and the scientists, on the other. In the process, I hope not only to emphasize several of the specific features of Russian and Soviet history, but also the specific features of fundamental science, a social institution quite unlike many others. And in the final section of the chapter, I maintain that the debates over science in Russia in recent years, so intense that they might actually be called "wars over science," tell us a few things about the nature of science that are helpful to Westerners as well as to citizens of the former Soviet Union.

The most common way to describe the fate of science in Russia (and subsequently in much of Eastern Europe) under Soviet rule is "dramatic political interference in academia."[1] According to this explanation, the Soviet system of organization and ideology of science was imposed by external political authorities upon unwilling and resistant scientists. No one who knows the history of science in the Soviet Union will deny that such political interference was a prevalent, constant, and harmful phenomenon. As we saw in chapter 3, the list of repressions is extensive, ranging from the trials of "bourgeois specialists" in the 1920's to the purge of the Academy of Sciences at the end of that decade; to the rise of Lysenkoism and other ideological perversions of science in the 1930's, 1940's, and early 1950's; to the persecution of dissidents and refuseniks in the 1960's and 1970's.[2]

However, an adequate description of science according to the Soviet model must be much more complex; along with resistance by scientists, there developed much collaboration and cooperation. Some aspects of the Soviet model of the organization and ideology of science were quite attractive to scientists: it gave them great prestige and a whole range of perquisites; the academy system largely freed the researchers from teaching duties; the same system gave senior scientists impressive institutional authority so long as they did not offer political resistance to the regime; and, finally, the ideology of Marxism gave "science" unique standing in the value system of society.

Only by seeing how attractive the Soviet model of science became to some scientists (particularly those who did not wish to emigrate) can we understand why "de-Sovietization" after the fall of the Soviet Union has been so difficult. In the Soviet Union, the old Academy of Sciences of the USSR (now the Russian Academy of Sciences) has proved to be one of the most conservative institutions in Russian society, mightily (and so far rather successfully) resisting post-Soviet major reforms. If the Soviet system was so repressive to science and learning, why do so many of the members of the Russian scientific establishment continue today, long after the collapse of the Soviet Union, to defend its major features so stoutly? In order to answer this question, we need to examine the history both of the establishment of the Soviet system of science after the 1917 Revolution and the fate of that system after the collapse of the Soviet regime.

From the very beginning of Soviet rule, the Party and government offered scientists a contradictory message, one that both repelled them and enticed them. On the one hand, most scientists found the Bolshevik rejection of Western standards of democracy and freedom to be unacceptable; to these scientists, the Bolsheviks were extremists, who, it might be hoped, would soon pass from the scene. In the 1920's, only a small handful of scientists and engineers belonged to the Communist Party. Not until 1929 was there a single member of the Academy of Sciences who was also a member of the Communist Party, and the first communists to enter the Academy were literally forced upon the establishment by political and police authorities.[3] On the other hand, the communists based themselves on a modernization program that gave science and technology prominent places, they called on scientists and engineers to help them in this effort, and they established scientific institutions in which scientists could play prominent roles as researchers and administrators. Despite their initial reservations about communist ideology, many scientists found it difficult to resist the allure of this scientistic program of modernization that offered so many influential positions to scientists and engineers willing to work with it.[4]

The interest of some Russian scientists in government-sponsored and government-controlled science actually arose long before 1917. Archival materials that have come to light in the past few years, since the fall of the Soviet Union, demonstrate just how strongly several influential administrators of Russian science agitated for such a system of scientific research during the last years of tsarism. One contemporary Russian historian of science recently observed: "Etatists by conviction, authoritative Russian scientists [V. I. Vernadsky, S. F. Ol'denburg, A. A. Shakhmatov] before 1917 worked out a conception and program of the *state* organization of science and became in the process—willingly or unwillingly—the founding fathers of the Soviet scientific-organizational system."[5] This quotation reveals the full poignancy of the situation: Vernadskii, Ol'denburg, and Shakhmatov are not usually remembered today as enthusiastic supporters of the Soviet regime; on the contrary, all three were members of the old prerevolutionary intelligentsia, opposed many aspects of the new Soviet order, and experienced political difficulties under Soviet rule. Yet all three also worked for the creation of a system of state-controlled research similar to what the Soviet regime favored. They believed that the state could offer support to science on a scale unprecedented in history, and they energetically favored the conversion of the Academy of Sciences from a rather small society of savants with almost no research facilities of its own into a vast network of research institutes, what Alexander Vucinich would later call an "empire of knowledge."[6] They actively worked for the creation of the golden cage that the Soviet system of scientific research subsequently became.

Many other scientists and engineers in Soviet Russia after 1917 harbored similar mixed feelings about the new regime, combining political suspicion with hopes of a rapid development of science and technology under Soviet sponsorship. In the immediate aftermath of the Bolshevik Revolution, in late 1917 and early 1918, political suspicion predominated, and many "bourgeois specialists" actively resisted the new regime, boycotting Bolshevik-dominated agencies and refusing to cooperate in Bolshevik campaigns. The Soviet authorities responded ruthlessly, and many arrests occurred in these early months. This period of resistance was, however, surprisingly brief. As Jeremy Azrael writes in his book *Managerial Power and Soviet Politics*, a "radical change occurred in the behavior of the 'bourgeois specialists' once resistance was suppressed. For, in a large and rapidly increasing number of cases, resistance appears to have given way not to passive compliance with orders but to conscientious collaboration and active cooperation. . . . Lenin heralded a 'huge reversal' in the outlook and behavior of the members of the technical intelligentsia as early as April 1919."[7]

The motivations of the scientists and engineers who moved toward cooperation with the Soviet regime were mixed. Many of them developed no conscious rationalization for their behavior beyond wanting to get on with their jobs; they would cooperate with any regime that permitted them to work in the specialties for which they were trained. To the extent that these specialists possessed an ideology, it could be called "professionalism": the belief that technical professionals should subscribe to neutrality in the political realm.

Other technical specialists accepted a broader concept of social and political responsibility and continued to nurture inner opposition to the regime, but developed a rationale that justified their temporary cooperation; people who subscribed to this rationale were sometimes labeled "the school of productive forces." According to this school of thought, the best way to achieve an eventual change in the Soviet regime, its evolution into an authentically socialist or democratic liberal order, was to do everything possible to facilitate economic development. Subscribers to this view believed that the technical and economic modernization of Soviet Russia would eventually lead to irresistible pressures from below for the amelioration of living standards and for the democratization of the political system. (In retrospect one could maintain that the supporters of this view were correct; however, they would undoubtedly have been appalled to learn how long and costly the evolution would be.) Quite a few scientists and engineers thought they could help this transition along its way, for they held influential positions in the educational, research, and economic establishments. Among the supporters of the "school of productive forces" were people of rather different political stripes. Some were proud Russian nationalists, believing that anything that made Russia stronger was good in the long run. Others took the same position for more liberal reasons, believing that economic development would inevitably lead to democratization. The liberal version was supported outside Russia by the émigré Constitutional Democrat N. V. Ustrialov, who in the early 1920's developed the doctrine of *smenovekhism*, which recognized the temporary stabilization of Bolshevism but called for its transformation from within by economic development.[8]

Yet a third position justifying cooperation with the Soviet regime by scientists and engineers was the one held by people who gradually internalized Soviet ideology and goals and who came to support the regime without question or reservation. Careerism certainly played a role here, as the expanding Soviet system of research, education, and industry opened up more and more positions for ambitious young specialists. And the prestige and perquisites that came with outstanding scientific achievement were greater than those in any Western society.

Although there were a number of positions that scientists in the USSR could adopt vis-à-vis the Soviet regime, they did not merely select from a menu of choices but actively helped shape the available alternatives. By working together with the government on such early modernization projects as the electrification project GOELRO, scientists and engineers helped elaborate Soviet policy in science and technology. The flow of influence between Soviet scientists and the regime was thus not unidirectional; negotiation might be a better word. If one can speak of the gradual "internalization of Soviet ideology" by some scientists, one can also speak of the gradual "scientization of Soviet ideology." Early Soviet Bolshevism had actually included some antiscientific elements, particularly in the proletarian culture movement, but gradually the ideology of the regime became more and more scientistic, an evolution aided by at least some of the scientists and engineers.[9] Eventually, the words *scientific* and *Marxist* became almost synonymous.

One of the aspects of Soviet science that is still incompletely understood is the role of dialectical materialism, what became the official Soviet philosophy of science. Today, most historians of Soviet science, both in the former Soviet Union and abroad, describe dialectical materialism as a dogma that was forced upon Soviet scientists. According to this interpretation, no Soviet scientists took the ideology seriously. As is the case with most other aspects of Soviet science, the situation was more complicated. With regard to Lysenkoism and a number of other major ideological incursions upon science, most scientists who were leaders in their fields were, indeed, horrified by the distortions of science that resulted from the imposition of particular versions of dialectical materialism by ideologists. But a full understanding of the role of ideology in Soviet science would also have to recognize, as I indicated in chapter 1, that Marxism was an important influence on some of the Soviet Union's best scientists—people who received international recognition for their work.

Science offered intellectuals in the Soviet Union a variety of relatively secure "niches." For those who increasingly accepted Soviet ideology and goals, it provided personally satisfying careers in which loyalty was rewarded. For those who still inwardly resisted this ideology, science provided relative freedom so long as they kept quiet about their inner convictions.

By the mid 1930s, Soviet science and technology had taken on the basic organizational and ideological features that remained constant down to the Gorbachev period, despite subsequent small changes. Through the 1960's and 1970's, the system continued to grow at a very impressive pace. Let us briefly take a look at that scientific establishment as it existed in the last years of the Soviet Union.[10]

At the beginning of 1991, the total number of people officially classified in the Soviet Union as scientific researchers and faculty was 1,520,000.[11] These scientists were distributed in three gigantic pyramids, which for sake of convenience can be called the "university system," the "industrial and defense ministry system," and the "Academy of Sciences system." The table in note 12 gives a very approximate description of the organization and shares of research personnel and budgetary funds of each of the three pyramids.[12]

From the standpoint of American experience, the most familiar pyramid of the three is the university system. The Soviet Union, like all modern countries, had large universities, all supported by the state in the Soviet Union, and the majority similarly supported in the United States. The superficial resemblance of the two systems should not blind one, however, to the enormous differences. In the United States, the universities are the home of most of the fundamental research conducted in the country. In the Soviet Union, the universities traditionally had a much narrower pedagogical role.[13]

The industrial and defense ministerial system was by far the largest of the pyramids. It was primarily concerned with applied science, although it performed some fundamental work as well (just as the Academy of Sciences system performed some applied work). One should note that most researchers in this industrial and defense system were not located in individual plants or "companies," but instead, like their colleagues in the academies of sciences, usually worked in centralized institutes in large cities. In the period of most intense growth of the Soviet scientific establishment (1960–72), the number of these institutes increased from 4,196 to 5,307.[14] After 1972, the number of institutes remained approximately stable, but the existing institutes continued to grow in size. The average institute in 1991 had a staff of 270 researchers, but some of the most important had research staffs of several thousand. The directors of these institutes were traditionally very powerful, even autocratic, figures.

The system headed by the Academy of Sciences of the USSR was the most unusual in terms of the experience of most Western countries. In many Western countries, the most prestigious learned academies and societies, such as the National Academy of Sciences in the United States or the Royal Society in Great Britain, are almost solely honorific organizations; they issue reports periodically and occasionally speak out on scientific issues, but they are neither the locus of laboratory research nor the place of employment for active researchers in the natural sciences. Members of these Western academies do not receive salaries because of their membership; in fact, they usually have to pay membership dues to the academies and derive their salaries from their other places of employment, often universities. The Soviet Academy of Sciences, on the other hand, was the major place of employment of the most outstanding fundamental researchers in the country. The same senior scientists who enjoyed the greatest honors and prestige the government could offer—often at rather advanced ages—were responsible for the actual administration, in the Academy system, of the research laboratories and institutes. Most full members of the Academy wanted institutes of their own, and it is not a coincidence that the total number of institutes in the all-union Academy was on the same order of magnitude as the total number of active full members. This was a system that tied official honor and administrative authority together to a remarkable degree. In my opinion, this combination undermines the productivity of science.

No official title or rank in Soviet society enjoyed greater prestige than that of *akademik,* full membership in the Academy of Sciences of the USSR. Anyone with this title could command an audience at meetings and publish articles in leading newspapers and journals without difficulty. And membership, or even corresponding membership, in this academy brought with it a heady array of privileges: special grocery stores, hospitals, sanitoria, limousines, housing, and foreign-travel privileges. The Academy controlled construction firms, thousands of apartments, kindergartens and schools, hotels, a university (Novosibirsk), several former estates of the nobility, a fleet of dozens of oceangoing ships, botanic gardens and nature preserves, and an array of consumer-service enterprises. It would be no exaggeration to say that the Academy controlled a branch of the national economy.[15]

Looking over this Soviet system of research, we can isolate some of its central features: it was large in numbers of researchers, highly centralized in organization, and dominated by powerful leaders. Fundamental researchers had little contact with applied researchers, and both fundamental and applied researchers had relatively little contact with students, especially undergraduates. Funding of research within this system was not dispersed on the basis of competitive applications by individual researchers, but by block funding of institutes by the central Academy presidium, which got its money from the government. This method of funding gave members of the presidium and institute directors great influence, the power of the purse. At the same time, block funding also provided some room for independent variations within these institutions, since scientists were not usually judged individually. This system tended to reinforce existing tendencies, usually mediocrity, but occasionally creativity and heterogeneity.

The system of Soviet research emphasized quantity over quality, seniority over creativity, military security over domestic welfare, and orthodoxy over freedom. The scientists who were in charge of this system had great authority, little accountability, few teaching responsibilities, and enormous personal privileges. No wonder they liked it.

Not surprisingly, this system of scientific research did not work as well as it should have in terms of scientific productivity. When one considers that the mature Soviet Union possessed the world's largest scientific establishment, one must conclude that the output of that system was disappointingly meager. No matter what criterion of excellence one chooses—number of Nobel Prizes awarded, frequency of citation of Soviet research, number of inventions registered abroad, or honorary membership in foreign scientific societies—the achievements of Soviet scientists were disproportionately small.[16] In the fields of natural science (physics, chemistry, physiology, and medicine) from 1901 through 1990, citizens of the United States received 145 Nobel Prizes and citizens of Russia and the USSR received 8. In all fields (including literature, economics, and peace), citizens of the United States won 217 Nobel awards in this period, and citizens of the USSR won 23. If one ranks the countries of the world in terms of the number of their citizens who received Nobel awards in this period, the Soviet Union ranked sixth, after the United States, Great Britain, France, Germany, and Sweden. And the Soviet Union seemed to slip badly in its last decades. The last Nobel award to go to a Soviet natural scientist was the physics award to Pyotr Kapitsa in 1976, and that was for work done in the 1930's.

Still, one must notice that Soviet science was also very strong in some areas. The Soviet Union was the first country in the world to build an atomic power plant, the first to launch an artificial satellite, the first to launch a human being into space, and the first to suggest the now widely accepted Tokamak model for nuclear fusion. In some areas of theoretical physics and mathematics, the Soviet Union was a world leader for decades, and it maintained strong traditions in oceanography, seismology, metallurgy, magnetohydrodynamics, chemical catalysis, and a variety of other fields. The major flaw of

Soviet science was not, then, its overall weakness. The major flaw of this science was the ratio of product received to effort expended. The Soviet Union devoted a larger share of its natural resources to science than any other leading industrial power, yet ranked below many of these powers in its scientific creativity and productivity.

If one looks at the sale of licenses for technology, the poor performance of Soviet research is particularly clear. The United States, in sharp competition with Western Europe, Japan, and other advanced industrial countries, sold approximately 30 times as many licenses for technology annually as the Soviet Union in the last years of its existence. A similar picture emerges from a study of honorary memberships in prestigious scientific societies. In 1986, shortly before the fall of the Soviet Union, the Royal Society of London had 87 foreign members, of whom 6 were citizens of the Soviet Union and 44 citizens of the United States. During the next two years, 3 of the Soviet members died, leaving only 3. The situation was slightly better among foreign members of the National Academy of Sciences in the United States, in which 16 members were Soviet citizens in 1986, out of a total of approximately 250 foreign associates.

Each criterion for comparing science in the Soviet Union with that in other countries has, of course, its flaws. The Russian language was known by few outside researchers, and consequently the achievements of Soviet researchers were more frequently overlooked than those presented in more accessible languages. It is hardly surprising that so many more American scientists were members of the Royal Society than Soviet scientists; the common language and many ties shared by British and U.S. scientists make such shared honors inevitable. Nonetheless, the conclusion is still inescapable that Soviet science and technology did not meet the grand goals that Soviet planners established.

This was not for lack of effort. As György Péteri has recently shown, the state-socialist countries suffered from "systemic over-stretch" in their scientific effort. The ratio of the density of research staff (full-time equivalent number of research workers per 10,000 inhabitants) against the level of economic development (GDP per capita) was far higher in most state-socialist countries than in capitalist ones.[17] Put simply, the state-socialist countries were trying harder in science than the rest of the world and achieving less. In terms of the national interests of the Soviet Union (and later Russia and the other republics), this system was one that cried out for reform.

The moment when reform seemed most possible came at the end of the 1980's and the beginning of the 1990's, as the Soviet Union collapsed. In the last months of the existence of the USSR, a strong reform movement developed there among some scientists, especially among young researchers in the fundamental sciences. These junior scientists were well aware of the deficiencies of the hierarchical Soviet scientific system, and they could not understand why a few hundred members of an honorary learned society, the Academy of Sciences, should be given administrative authority over all fundamental science and virtual ownership of a branch of the national economy. Most of these critics of the existing scientific establishment also favored democratic reform of the Russian government. In fact, they were far more unanimous in their opinions about how the *government* should be organized than they were in their views about how *science* should be organized. As a result, they were never able to form an effective political coalition on matters of science policy. They did not have clear answers to crucial questions such as: "As the Soviet Union makes the transition from totalitarianism to democracy, how much of the scientific establishment needs to be changed in step with the political changes sweeping the country?" "To what degree did science in the former Soviet Union display the characteristics of the centrally controlled political and economic system that had formed and nurtured it?"

The reformers split into different groups in their answers to these questions. Some of them favored what was often called "the American system," which meant that they believed that the seat of fundamental science should be in the universities, not in institutes of the Academy. Others believed that preservation of the institute system was necessary, but favored separating these institutes from the Academy of Sciences, which would become a mere honorary society without administrative authority. In this scheme, the institutes that had previously been in the Academy would be reorganized into an association of research laboratories, perhaps like the Centre National de la Recherche Scientifique (C.N.R.S.) in France or the Max Planck Gesellschaft in Germany. Still others favored keeping the basic organizational structure of the old Soviet system of science but proposed to "democratize" it by including in the major decision-making bodies of the Academy—the General Assembly and the Presidium—elected representatives of the scientists throughout the Academy, including the junior scientists. Opposing all three of these blueprints for the future were the top officials of the old Academy of Sciences of the USSR, who struggled mightily almost up to the collapse of the Soviet Union itself to keep the old all-union structures, along, of course, with their offices and privileges. They pointed out that whatever the flaws of the Academy of Sciences of the USSR, it nonetheless possessed the best scientists in the country. They feared that a radical reform of Russian science would destroy its strongest institutions.

At the peak of the wars over science, which came in the period 1989–91, many different positions emerged. Everybody, it seemed, had grievances about science. Pure scientists, researchers in fundamental areas, chafed at governmental and political controls and at the preferences given to topics of industrial and military importance. They wanted more money for fundamental science and less interference.[18] Soviet government bureaucrats and economic planners from GOSPLAN considered the pure scientists unaccountable and arrogant; these governmental administrators pointed out that the USSR spent a higher percentage of its gross national product on science than any other country, in fact, more than all West European countries taken together, and one and a half times as much as Japan. Yet Soviet scientists were far less productive than Western and Japanese researchers; according to one GOSPLAN official, the average Soviet scientist was four times less productive than the average American scientist.[19]

Scientific workers in the Academy of Sciences, especially the junior ones, complained that they were entirely excluded from the influential policy-making bodies of the Academy. They called for a more democratic system of administration within the Academy.[20] Corresponding members of the Academy of Sciences resented their junior status and wished to be full members; they pointed out that in other countries, a two-rank system of membership in academies and scholarly societies was rare.[21] Engineers resented their expulsion from the Academy in the 1960's and called for a new Academy that would include them.[22] University professors were unhappy about the lower status of university science in relation to that in the Academy and called for a transfer of science to the universities; if such a transfer were not possible, they wanted the rank of "academician" for themselves.[23] Russian nationalists were unhappy over the absence, in the Soviet system, of a Russian Academy of Sciences, and asked why Russia was the only republic in the union without its own academy of sciences.[24] Political reporters and outside observers were critical of the arrogance and philistinism of the leading administrators of the Academy of Sciences of the USSR. A particularly offensive symbol of their expensive bad taste was, in the critics' opinion, the garish new skyscraper intended for the presidium being constructed near Gagarin Square; this building would require a service staff of 600 people, and its offices were originally slated to be entirely those of administrators. In the revolutionary situation that was developing in science, the building served the role of Marie Antoinette's diamond necklace before the French Revolution, inflaming opinion against the old regime.[25] The deputies in the increasingly democratic parliaments of Russia resented the financial unaccountability of the Academy of Sciences to the government; the Academy always wanted more money, and it wanted absolutely no interference in how it spent the money. The ire of the deputies was inflamed by Mikhail Gorbachev's August 1990 decree making the Academy fully independent of government influence, and able to possess its own property, while remaining totally dependent on the government for its budget. (This decree was later reversed by Boris Yeltsin.)[26] Provincial scientists, those outside Moscow, St. Petersburg, and Novosibirsk, resented the "big science" centered in those cities. They wanted the status awarded scientists in the Academy, most of whom were in Moscow. These provincial scientists were the driving force behind the creation of a rival Russian Academy of Sciences in 1990 and 1991, and they also created a dozen or more other rival "academies."[27]

The position of the old guard, the supporters of the status quo, was enormously weakened by the events surrounding the abortive putsch in August 1991. When army generals and police officials attempted to reverse Gorbachev's reforms and restore authoritarian, if not totalitarian, rule, the top leadership of the Academy of Sciences silently sided with them. The Academy's leaders were, in fact, along with the conservative heads of the Writers' Union, among the very few intellectuals in the dying Soviet Union who sided with the old order. The leaders of the Academy of Sciences of the Soviet Union believed that their fate was inextricably tied to that of the USSR itself. When the coup failed and its leaders were arrested, the leadership of the Academy was thoroughly discredited. *This* was the moment when reform was possible.

In the immediate aftermath of the coup, the old Academy was in a very precarious position. The Soviet Union was breaking up into its constituent republics, and already a rival "Russian Academy of Sciences" had arisen. In its initial conception, this new academy allowed no place for the old system; according to the "basic principles" of the new academy published in the press: "The Academy of Sciences . . . does not contain scientific-research subdivisions or schools" and "Material compensation for membership in the Russian Academy of Sciences is not provided."[28] If this model had been followed, a radical reform of Russian science would have occurred, moving Russia toward either a university-based research system or an institute-based one that was separate from the Academy of Sciences. According to these first principles, the new Russian Academy of Sciences would become a mere learned society, prestigious but not administratively important. However, this transition did not occur. As Aleksei Zakharov, a leader of the reform movement observed, "An historic opportunity for a constructive reformation of the entire organization of Russian fundamental science was missed."[29] What happened?

What happened was that the allurements of the old system were too enticing even for many of the reformers to resist. As Zakharov observed, the people organizing the new Russian Academy of Sciences "faced a choice: lead an academy without privileges and institutes, or be rank-and-file members of an academy with institutes, salary supplements and privileges. They chose the second option. Which would you choose?"[30] In this stark question we see the error in the position of those people who maintain that the Soviet system of scientific research was imposed on unwilling scientists. Many of those scientists had come to *prefer* the old Soviet system.

A way to make the transition to a new independent Russia without giving up the state-socialist system of fundamental research was found by fusing the two academies, the old Academy of Sciences of the USSR and the new Russian Academy of Sciences; the approximately 300 full members and 450 corresponding members of the "big" Academy were combined in late 1991 with the newly elected 39 full members and 108 corresponding members of the just-born one to form a single "Russian Academy of Sciences" (RAN), which took over the research establishment of the old Academy. A part of this agreement was the abandonment of the idea of true reform that had motivated many of the early advocates of the Russian Academy of Sciences. Instead of becoming a learned society in the West European or North American mode, the Russian Academy of Sciences remained a vast research bureaucracy of laboratory and institutes claiming primacy in fundamental research. Thus, for the foreseeable future, Russia will have a scientific establishment that is distinctly different in its organizational principles from the scientific establishments of most countries in Western Europe and North America, and most similar to that of the old Soviet Union. And I predict that this system will continue to suffer from the maladies of inefficiency and low creativity that plagued the old system.

To be sure, some reforms have been enacted. The main decision-making body of the Academy, the General Assembly, now contains elected representatives of the institutes, and therefore the rule of the full members is no longer absolute, although their dominance does not seem to be threatened. The method of financing research has been reformed, with research proposals and peer review replacing, to a certain extent, the old system of block funding. A government foundation, the Russian Foundation for Fundamental Research (RFFI), has been established that awards grants to principal investigators on a competitive basis. This change in the financing of research is potentially very important, but it has been undermined by the fact that the Russian government has not been able to assign large funds to the foundation and also by the fact that the system seems to be working within the tradition of favoritism and even corruption that plagues much of the Russian economy; the scientific advisors to the RFFI, senior scientists, seem to have been among the primary beneficiaries of its grants.[31]

Conclusions

Why has the scientific establishment of Russia, particularly in fundamental science, been so resistant to change? Several different factors have influenced this process:

1. Support for the old system among quite a few scientists, particularly the senior ones, who benefited enormously from it.

2. The dependence of fundamental science upon the state

budget. Fundamental science cannot exist on a self-sustaining basis, and so far no alternatives to state support, such as grants from private foundations or industry, have arisen in Russia (with the exception of grants from foreign foundations, which are probably a temporary phenomenon). The old state-socialist model of science is now seen by many scientists, young and old, as their only hope for protection from market forces.

3. A widely shared belief among scientists and government leaders in Russia that—whatever the flaws of Soviet science—science and science education were actually among the strongest features of the Soviet Union, and that it would be foolish to change the system radically.

4. Surprisingly strong sympathy among scientists for some aspects of Soviet ideology, especially its emphasis on science as the most important aspect of culture, but even, to some still undetermined degree, dialectical materialism.

5. A lack of unanimity—not only in Russia, but throughout the world—about what kind of organizational forms are best for science. Scientific research is organized rather differently in leading industrial nations—such as France, Germany, the United Kingdom, the United States, and Japan—and no one knows what "the best system" is. Such uncertainty leads countries with large scientific establishments that have worked fairly well—and the Soviet Union probably fits that description—to tinker with their existing systems rather than radically reform them.

6. An economic crisis that meant that there was simply no money to finance major reforms in Russian science, such as a substantial beefing up of university research, a reform that almost everyone agrees would be beneficial.

In many ways, the story of the abortive reform of Russian science seems strangely and essentially *Russian*, without much reference to more general issues of the governance of scientific institutions, the sort of issues that Westerners worry about in their own countries. But underneath the debates about the organization of science in Russia are three questions of universal importance, which, unfortunately, have never been sufficiently disentangled and separated: (1) How should a democratic country govern itself and how should it choose its parliamentary representatives? (2) How should academic research institutions in a democratic country be governed? (3) How should explicitly elitist honorary societies choose their members and govern themselves in a democratic society?

In Russia at the peak of the debates, these three questions were thoroughly and confusingly intertwined. At the end of the Soviet period, the Academy of Sciences was held responsible for answering all three of these questions. As one of the "social organizations" allocated seats in the Congress of People's Deputies, the parliament, the Academy was responsible for electing representatives to the legislature for the entire Soviet Union. As the institution responsible for fundamental research, the Academy also had the obligation to govern the leading scientific research institutions in the country. And as the country's most prestigious honorary society, the Academy was responsible for choosing the small elite of intellectuals receiving the highest accolades the country could bestow. The failure to distinguish these three different functions and to see that different principles of governance are appropriate for each explains much of the heat and confusion found in the debates.

The moment when this confusion was greatest was in February 1989, when the Academy was trying to send its representatives to the Congress of People's Deputies. This purely political function had nothing to do with science itself. According to the election rules, the Academy was allocated 25 seats in the parliament. The attempt by the ruling presidium of the Academy to appoint these individuals according to the rules by which it did all its other business—the administrative-command method, with no pretense of democracy—inflamed the situation. When the presidium refused to put Andrei Sakharov and Roald Sagdeev forward as deputies, despite their popularity with the rank-and-file scientific workers, the lower ranks of the Academy researchers revolted. They assembled in a large demonstration in front of the Presidium building of the Academy and forced the administrators to back down and nominate Sakharov and Sagdeev.

Once the arbitrariness of the leadership of the Academy had been fully revealed by the "Sakharov and Sagdeev Affair," the demonstrators moved on to other issues, demanding reforms in the way in which the presidium governed scientific research and even in the ways in which the Academy introduced new members. It was easy for the reformers to point to many instances of corruption in the past practices of the Academy administration and in the election of new members.[32] But on issues directly related to scientific research, the demonstrators were on weaker ground; even in the most democratic countries in the world, scientific institutions are not governed democratically, and in no country are the members of honorary societies chosen democratically. In the governance of scientific institutions, practices vary widely internationally, but usually administrators of such institutions are chosen from above, with some consultation of the rank and file. In admission to honorary societies, the practice is overtly elitist in all countries, with the members of these societies producing their own successors, usually by secret ballot. The administrators of the Academy fought a long and hard battle against the reformers on these issues and eventually won.

One of the arguments that the administrators of the Academy used in these debates was to point out that scientific institutions are not governed democratically in the most advanced Western societies, and that members of honorary societies are also not chosen democratically. Thus, they were saying, "We are just doing the same things that other countries do, even the ones that have long democratic traditions." What the top leadership of the Russian Academy of Sciences failed to say, however, was that in no other country in the world were the functions of scientific administration and honor societies *combined.* The members of the Royal Society in London and the members of the National Academy of Sciences in Washington have *no* administrative functions in scientific research determined by their membership in these societies. The Royal Society in Britain and the National Academy in the United States have no research laboratories of their own in the natural sciences; they serve a purely honorific and advisory function. In most Western countries, the various functions discussed above have been clearly separated. In Russia, these functions have remained intertwined to the present day, just as they were in the old Soviet Union. The members of the Russian Academy of Sciences, chosen as "immortals" for their scientific distinction, are also responsible for administering the research institutions of fundamental science.

Thus, years after the collapse of the Soviet Union, Russia remains saddled with a system of scientific research that is more similar to its Soviet predecessor than perhaps any other aspect of Russian society.

CHAPTER FIVE

Who Should Control Technology?

The technology projects that societies embark upon in the course of modernization vary greatly in their size and in the degree to which they impinge upon the lives of their citizens. Within the history of industrialization of each nation, a few projects stand out by virtue of their size, expense, and massive social effects. Examples of such exceptionally large and socially disruptive projects include the building of canals and railroads in nineteenth-century America and mammoth hydroelectric power plants, flood control systems, reservoirs, and, since World War II, the interstate highway system. For the purpose of convenience, I shall call these enormous projects, costing great amounts of money and involving the lives of tens of thousands of citizens, "megaprojects." Many, but not all, U.S. megaprojects have been "public works" projects in the sense that they have been financed by federal, state, and local governments. Others, such as railroads and great airports, have been financed privately (often with government support and incentives) or by a combination of governments and private industries.

In the Soviet Union and in other Soviet-style economies, such as those of China and Eastern Europe until recently, all megaprojects have been planned and financed by the government. In the Soviet Union during the early five year plans, many of these construction projects were known as *stroiki* and were a part of the romantic mythology of Soviet industrialization. They included Dneprostroi, the building of what was at the time the world's largest hydroelectric power plant; Belomorstroi, the White Sea Canal connecting the Baltic and White seas; Magnitostroi, the construction of what was and still is the world's largest steel plant; and Kuznetstroi, the development of the coal fields and railroads needed to supply coke to the blast furnaces at Magnitostroi. After World War II, one of the most massive megaprojects in the Soviet Union was BAM, the two-thousand mile Baikal-Amur railway, built through some of the most difficult terrain in the world. The Soviet Atomic Project and the Soviet Space Program were also megaprojects, demanding the participation of hundreds of thousands of people. Indeed, when we remember the "gigantomania" that characterized Soviet industrialization, we may be tempted to say that "megaprojects" or "megaplanning" characterized the Soviet economy more than any other in the world. Several of these projects, such as the "Northern Rivers" plan to divert the flow of Siberian rivers to the south, were never actually built, although they were much discussed. The Soviet Union on several occasions exported its approach to megaprojects to other countries. Perhaps the most graphic example is the Aswan Dam in Egypt, built largely by Soviet engineers.

During the twentieth century, the attitudes both of the larger public and of technical specialists in the Western world to such megaprojects changed in a fundamental and irreversible way. In the early part of this century, the view was widespread that such projects should be entirely planned and executed by technical specialists. Industrialization, construction, and modernization were activities about which specialists—industrial managers, engineers, economists, government planners—were thought to possess superior knowledge. This technocratic ideology was widespread through all industrialized nations, and was particularly strong in the United States, where the idea of progress was often linked to the advance of technology.[1] In the United States at the end of the nineteenth and beginning of the twentieth centuries, the doctrine of progressivism included the belief that good public works projects could be built and operated efficiently only if technology was divorced from politics and placed under the control of engineers.[2] As the American historian Robert Wiebe observed, Americans in the early twentieth century had a childlike faith that engineers were public-spirited experts, immune to graft and corruption, who could provide answers to all types of problems.[3] Herbert Hoover, who was elected president of the United States in 1928, was popularly known as the Great Engineer and had won his fame by launching such enormous public works projects as the Boulder (later Hoover) Dam on the Colorado River. It was the Soviet Union, however, that was to take to its apogee the idea of the planning of megaprojects by technical experts with minimal input from the public.

But just as the United States was a place where the progressive idea that engineers might be excellent social planners achieved early strength, it was also the place where the technocratic dream faltered most dramatically. The loss of faith in technical experts by Americans was a long process, which I do not have space here to describe in detail, but perhaps the most traumatic moment in that history was the "Freeway Revolt" of the late 1960's and early 1970's, when the interstate highway construction project—the largest public works project in history—was brought to a halt in a number of cities, including San Francisco, New Orleans, Boston, Philadelphia, and San

Antonio. The public had been alienated by the high-handedness of the engineers, who were ramming multilane highways through city neighborhoods, uprooting thousands of citizens, and destroying entire communities. The rallying cry of the rebels was that the public had been almost completely excluded from the planning process. In other words, the motto of progressivism at the beginning of the century was turned on its head: the point was not to exclude politics from technical planning but to make it an integral part of that planning.

Against the background of this evolution of attitudes in the United States toward the planning of megaprojects, the history of Soviet industrialization is particularly interesting. In the Soviet Union, the tendency toward the technocratic planning of megaprojects that had once been so characteristic of the United States continued to develop into a particularly monstrous form, which totally disregarded the public. American-style affection for megaprojects planned by engineers was wedded in the USSR to the scientism of Marxism and the authoritarianism endemic in Russian politics to form an almost unstoppable oligarchy of technically trained industrial planners. Only the demise of the Soviet Union could halt this monstrous technocracy. Even now, Soviet-style technocracy lives on in China, where the Three Gorges Hydroelectric Project is being pushed in the same way that Soviet ventures once were—that is, by the total exclusion of the public. Indeed, the leading exponent of the Three Gorges Project, Premier Li Peng, is a graduate of a Soviet engineering institute of the 1950's, and he still espouses an ideology typical of Soviet industrial planners.

Today, with our knowledge of how catastrophically the Soviet Union failed, it is difficult to recapture how fascinated the West was in the 1920's with the Soviet effort to industrialize Russia. Few engineers and technical specialists in the West were attracted to Marxism, but many of them were intrigued by the possibility that the Soviet Union might catapult technocrats into positions of power and might create a society based on technical rationalism. In his 1921 book *The Engineers and the Price System*, Thorstein Veblen called for the creation of committees he called "soviets" (the influence here of the Russian Revolution is unmistakable) to run the economy in a "rational" way.[4] The 1920's was a period when engineers in the United States enjoyed a popularity never again to be repeated.

The American engineer who was perhaps most outspoken in the late 1920's and early 1930's in calling for cooperation with the Soviet Union was Hugh Lincoln Cooper (1865–1937), who had gained much experience working for the U.S. Army Corps of Engineers and later established his own consulting company for the construction of hydroelectric dams and power companies.[5] Cooper, a member of the Republican Party and a defender of capitalism, seems an unlikely person to have been a partner with the communists of Soviet Russia, but his commitment to business and construction overcame all ideological reluctance. In September 1928, Cooper spent two days with Stalin in the Caucasus discussing industrialization. Cooper later told an American audience that the Soviet Union was "the world's greatest human experiment"; he added that he had been "greatly impressed" by Stalin's "broad grasp and keen analysis of the problems in Russia." "Stalin," he continued, was "kindly minded but firm and confident that their economic plans are correct."[6]

In the early 1920's, Cooper had been involved in the construction of the first dam in what was to become the TVA, or Tennessee Valley Authority. In his accounts of his work, Cooper paid little attention to the families of the river valleys in Tennessee who were dispossessed and removed prior to the construction of the reservoirs. In the United States, the dispossession and removal was accomplished by the legal doctrine of eminent domain, and the farmers whose land was taken were paid for their land. Frequently, the subsistence farmers were happy enough to take the money and leave; in other cases, they resisted. Of the 3,500 families removed for the Norris Dam basin, 96 refused to leave voluntarily and had to be forcibly evicted.[7] Some of these families had owned land there for four or five generations. The TVA was built according to the progressive ethos that specialists know best and that the government has the right to remove large numbers of people when it thinks it proper to do so.

In 1926, the Soviet government asked Hugh Cooper to act as a consulting engineer for the construction of a gigantic hydroelectric power plant on the Dnieper River, what was to become Dneprostroi. Cooper began work in May 1927. He observed that the Russians "are looking particularly to America now for guidance. . . . I believe we can extend this aid without departing from or sacrificing any of our own ideals."[8] In other words, the politics could be kept out of the technology.

To clear the area for the reservoir, over 10,000 villagers were forced out of their homes, many of them German Mennonites.[9] A strongly religious people with above-average wealth in land and livestock, the Mennonities were seen as ideological opponents to the Soviet order.[10] Considered kulaks, they were not compensated for their land. Some of the Mennonites made the transition well, accepting what they saw as God's fate. Others resisted, were arrested, and became a part of the forced labor contingent at the work site. These prison laborers were marched to work under armed guard and were given the most difficult jobs.

Initially, the builders of Dneprostroi promised to provide

adequate housing and cultural facilities for the workers. As the project proceeded, falling behind in its schedule and grossly exceeding the estimated costs, the workers' needs were more and more neglected. By the time the project reached full speed, the workers were living and laboring under miserable conditions. As Anne Rassweiler wrote in her history of Dneprostroi, "Barrack dwellers complained of snow drifting through rooms. Tent dwellers endured temperatures below -13 degrees Celsius in the winter, and tornado-strength winds whipped tents away in the summer of 1929. Crowding, dark and noise were endemic. Toilet facilities were inadequate and frozen in the winter months."[11] In the following months, conditions became worse, and food grew increasingly scarce. Flour had to be delivered to the bakeries at night under armed guard to prevent theft. The poor nutrition contributed to outbreaks of disease. Tuberculosis, typhus, typhoid fever, and smallpox spread throughout the barracks and tents. No one knows to this day how many people died.

Hugh Cooper was later asked if it was true that forced labor was being used at Dneprostroi and other construction projects in the Soviet Union. He replied that he had traveled all over Russia and Siberia and had not once seen any signs of involuntary labor. Such charges he said, are "propaganda . . . made of whole cloth."[12] As a result of memoir literature recently published in the former Soviet Union, we now know that tens of thousands of prisoners were used at Dneprostroi, Magnitostroi, and Belomorstroi.[13] Many of them were peasants dispossessed by collectivization, an operation described by Cooper in 1931 as "rapid progress" that "now embraces over 60 percent of all peasant households" and that produced harvests that were "the largest in Russian history."[14] We now know that collectivization produced an enormous famine in Ukraine, in which millions died.[15]

Ironically, the most astute critics of the early Soviet industrialization projects were not capitalist engineers from the West such as Cooper, who were eager to make money off the Soviet projects, but socialist engineers from the Soviet Union itself, headed by the chairman of the Russian Technical Society, Petr Pal'chinskii.[16] Pal'chinskii was a consultant for a number of the early *stroiki,* including Dneprostroi and Magnitostroi, and he warned the Soviet government not to go ahead with these projects until a more thorough evaluation had been made of their technical characteristics and their social consequences. Pal'chinskii believed that socialist Russia had the opportunity to develop a far more humane industry than anywhere else. While Pal'chinskii admired American engineers, he thought that they were too narrowly interested in technology, ignoring its social context and its social consequences. He called for "humanitarian engineering" as a substitute for the American doctrines of Taylorism and Fordism.[17] The single most important factor in engineering decisions, Pal'chinskii maintained, was human beings.[18] Successful industrialization and high productivity were not possible, he repeatedly emphasized, without highly trained workers and adequate provision for their social and economic needs. The Soviet Union, he warned, must have a goal beyond the construction of heavy industry. It must also aspire to a society where all human needs were fulfilled, and where industry was made to serve the needs of the citizenry, rather than the reverse.

Turning to the project for the construction of the world's largest hydroelectric power plant at Dneprostroi, to which Hugh Cooper had been invited as a major consultant by the

Soviet government, Pal'chinskii was horrified by the haste, negligence, and inhumanity with which the dam was being built. He pointed out that the plans for Dneprostroi were going ahead before complete geological, hydrological, or topographical maps had been prepared. As a result no one knew just how large an area the dam, which was to be 35 meters (96 feet) high, would flood. Furthermore, no one had done studies of the alternatives to the giant dam that were available, such as a series of smaller dams, or the construction of thermal power plants that would use the locally available coal. Finally, Pal'chinskii pointed out that the dam would force the evacuation of 10,000 villagers from their homes. He noted that no one had included the loss of these farmlands in the estimates of the expenses of Dneprostroi; nor had anyone considered the human costs in forcing these people off the land.

Pal'chinskii advanced a similar criticism of the Soviet plans to build the world's largest steel mill at Magnitogorsk.[19] He criticized the plan for its negligence of the human factor, pointing out that there were no population centers nearby, and that an entire city would therefore have to be built for the labor force. He reminded the Soviet government that adequate workers' housing and urban amenities were preconditions for such a gigantic complex. Pal'chinskii doubted that the Soviet government was making adequate provisions for the residents of this new city. Pal'chinskii also noted that although everyone marveled over the rich ore of the Magnetic Mountain, no one had yet made a thorough study of how much it contained. It was quite possible that after a few decades, the ore would be exhausted, while the continued presence of the gigantic steel mill would require the costly hauling of ore from other regions. He noted that no coal was available near the projected city of Magnitogorsk, so that from the very beginning, fuel for the voracious blast furnaces would have to be hauled in by railway, although water transport is by far the least expensive means of hauling such heavy loads as iron ore and coal. He pointed out that in the United States, the steel mills are located, not near the rich ore deposits of the Mesabi Range in Minnesota or the Marquette Range in Michigan, but hundreds of miles away in Detroit, Gary, Cleveland, and Pittsburgh—all cities with large labor forces, the first three served by water transport, the last near enormous coal deposits.[20]

Pal'chinskii's worries about Magnitogorsk were ignored. The Soviet government announced that the new steel complex would be equipped with the latest technology and would surpass all Western competitors in size and quality. The last word in steelmaking at the time was the United States Steel works in Gary, Indiana, and the Magnitogorsk mills were to be bigger and better. To fulfill this goal, the government brought American engineers, some of them from Gary, to help plan the mills. The foreign engineers were not asked whether Magnitogorsk was the right spot to build such a mill; they were simply told to advise on its construction. Trained in narrow technological terms themselves, the American engineers seemed willing enough to comply. The American engineers were housed near the planned mill in a special settlement called by them "Amerikanka," where they enjoyed luxuries not available to their Soviet counterparts, such as private houses with modern conveniences and even tennis courts.

The 200,000 workers who were brought to the city lived meanwhile mostly in barracks, tents, and mud huts in conditions of filth and deprivation, without indoor plumbing and surrounded by open sewers. Many of the workers' barracks were directly in the path of the fumes from the blast furnaces.

Much worse than the regular workers' lives at Magnitogorsk

were those of the 30,000 or so kulaks, peasants dispossessed by the collectivization of agriculture and brought as prisoners hundreds of miles to work on the construction of the steel city. Laboring under armed guard, they did the heaviest and most disagreeable work. Without adequate food or clothing, and housed in tents, an estimated 10 percent of them died during the first winter. Later they were placed in barracks that housed 40 to 50 families each. These barracks remained the domiciles of the "former kulaks" and their children for decades; only in the late 1960's and early 1970's were the survivors able to move to apartment buildings. Even as late as 1989, 20 percent of the apartments in Magnitogorsk were of the communal type, in which each bedroom was occupied by a separate family, and the kitchen and bathroom were shared by all.[21]

Pal'chinskii voiced the suspicion in 1927 and 1928 that the inadequately explored iron deposits of the Magnetic Mountain would be more limited than expected and the location of the giant steel mills away from waterways might turn out to be unwise.[22] By the late 1960's, that skepticism proved correct; the exhaustion of local iron ore required the mills to import ore by train from other regions, as they had from the very beginning the coal for the blast furnaces. From this time on, Magnitogorsk had no rationale whatsoever for being where it was, and its managers were forced to obtain both of its heavy raw materials from a great distance over land, rather than by cheaper water transport. Magnitogorsk, the world's largest steel mill even today, has become a monument to inefficiency.

In 1987, the young historian Stephen Kotkin from Princeton University spent several months in Magnitogorsk, the first American to live there for any length of time since the early 1930s.[23] He found a dirty and dispirited city surrounding hopelessly obsolescent steel mills. He observed "a crumbling or nonexistent urban infrastructure, almost unfathomable pollution, and a health catastrophe impossible to exaggerate."[24] As Pal'chinskii had predicted, the Soviet Union had created a monstrosity by ignoring the social context of engineering. Foreign engineers willingly participated in the creation of this hyperbolic example of socially insensitive technology; they were, if anything, even less socially aware than their Soviet colleagues.

Pal'chinskii's warnings about industrialization projects that ignored social context and social impact were ignored. He was arrested, accused of anti-Soviet conspiracy as the leader of the "Industrial Party," and executed without trial. Only with the opening of the Soviet archives with the collapse of the Soviet Union have we been able to examine his papers and read his prophetic warnings.[25]

From the time of his death at the end of the 1920's until the collapse of the USSR, Soviet planners continued to industrialize with little consideration for social or environmental consequences. Soviet engineering students were given extremely narrow educations, with no humanities or social sciences courses except three required ones on Marxism and the history of the Communist Party. In the last decades of the Soviet Union, these narrow technocrats came to dominate the top leadership of the country. Between 1956 and 1986, the percentage of members of the Politburo who had received their educations in technical areas rose from 59 percent to 89 percent.[26] These were men who favored heavy industry above all other priorities. We all know now how public health and nutrition suffered, and that the environment around industrial sites was a disaster.

At the present time, two of the largest public-works projects in the world, and perhaps the very largest, are the Three Gorges Hydroelectric Plant on the Yangtze River in China, and the Central Artery/Harbor Tunnel Project in Boston, Massachusetts. A brief comparison of these two megaprojects will illustrate that they are being pursued and promoted in vastly different ways. They are products of radically different paradigms about the relationship of massive technology to the society in which it is situated. The Three Gorges Project is a surviving relic of Soviet-style planning from above in which closed decision-making by political and technical elites still governs; the Central Artery/Harbor Tunnel Project in Boston is a product of a new approach to public works in which community involvement is a prerequisite and in which social assent is considered as important as technical competence. Each of these projects has a long history, and each today bears the marks of that history.

The idea of building a massive dam on the Yangtze River has been around since the 1920's. Even long before that, many of China's greatest emperors saw the construction of dams and canals as their central roles.[27] Both Sun Yat-sen, founder of modern China, and Mao Zedong, creator of the People's Republic, strongly favored the Three Gorges Project. The dam would be the largest construction project in China since the building of the Great Wall itself.[28] Present cost estimates conclude that the price of the project will be between U.S.$17 and 30 billion.[29] It will displace 1.4 million people and flood more than 100 towns, 800 villages, and nearly 100,000 hectares of China's best farmland.[30] The reservoir created by the dam will stretch for over 250 miles. Proponents of the dam maintain that it will control the floods that periodically hit the Yangtze river basin and will supply electricity for Chinese industry.[31] Critics point to the fact that the project will displace hundreds of thousands of families from their ancestral homes, create a huge number of refugees, and cause uncalculated environmental damage.[32] They also emphasize the potentially disastrous effects of siltation and earthquakes.

The Three Gorges Dam in China not only represents a survival of Soviet-style industrial planning but is organically linked to the now-defunct old Soviet Union. In the 1950's, when relations between China and the Soviet Union were close, a team of several dozen Soviet experts worked with Chinese engineers in developing the master plan for the siting of the dam and the reservoir.[33] As already noted, the most ardent supporter of the Three Gorges Dam today is Li Peng, a former power industry minister, who is a great exponent of megaprojects of the sort so long favored in the Soviet Union. But the favor given such projects in China is not just a result of Li Peng. The top leadership of the Chinese Communist Party is dominated by engineers, as was the case in the last years of the Soviet Union. As of August 1, 1993, of the nineteen members of the Politburo of the Chinese Communist Party, fourteen had received higher education, and eleven of these were engineers.[34]

The Three Gorges Dam project has been controversial, but most of the controversy has not been between the centralized planners and the public, but among the various ministries and bureaucracies that are interested in the project. With only a few exceptions—to be mentioned further along—the Chinese government has squelched public controversy through a system of censorship and political control. As a result, most of the decades-long discussion has not been about whether the dam should be built but about how and when. Bureaucrats interested in flood control have generally favored a higher dam; those in charge of the generation of electrical power have often called for a lower dam or a series of smaller ones. This debate among technocrats has not revolved around the most serious issues raised by the dam—namely, the dangers it presents to the environment and the society.

Only in the past ten years has the debate over the Three Gorges Dam reached the public arena, and then it did so only briefly. The course of that flickering public debate is intimately tied to the fate of Chinese dissent, and it tells us much about Chinese politics.

In 1988 and 1989, an extraordinary event occurred in Chinese politics. A group of prominent artists, journalists, scientists, and intellectuals organized independently of the government and launched a protest campaign against the Three Gorges Dam.[35] They published a book of interviews and critical essays about the dam entitled *Yangtze! Yangtze!* that became a watershed in the aborning Chinese dissident movement.[36] One of the editors of the book, Chengjing Jie, called a press conference at which he declared: "We hope the authorities will halt this big-name, big-money, low-benefit project that serves as a monument to a handful of people." A memorandum prepared by the dissidents further observed, "For the first time ordinary Chinese people will not keep silent on a weighty economic policy decision."[37]

At first it appeared that this wave of protest might actually achieve the goal of stopping the dam. At the 1989 National People's Congress, hundreds of delegates called for the postponement of the construction of the dam until the next century. A few days, later the government announced that the project would be shelved for five years.[38]

The sinologist Frederic Moritz of Pennsylvania State observed that Premier Li Peng's loss of face over the Three Gorges Project helped stimulate the student protests for democracy in Tiananmen Square.[39] By stopping the dam, or by at least postponing it, the burgeoning democracy movement gained confidence that it could change the Chinese political system. Their hopes were not realized.

After the brutal suppression of the students on Tiananmen Square in 1989, Dai Qing, the chief editor of *Yangtze! Yangtze!* was arrested and jailed without trial for ten months.[40] *Yangtze! Yangtze!* was banned, and criticism of the Three Gorges dam was forbidden, as it still is today. The Chinese government, led by its premier, the Soviet-trained engineer Li Peng, continued to push for the construction of the dam.

Despite the suppression of dissent, the dam continued to be controversial. At the March 1992 session of the National People's Congress, the chairman refused to allow discussion of the project.[41] In protest, several delegates stood and shouted their opposition. The government rammed through a vote in favor of the project, but for the first time in the history of communist-led China, one-third of the delegates registered their opposition to a government proposal by voting no or abstaining. But two-thirds voted in favor, and the Three Gorges dam was officially approved.

Since then the Chinese government has tried to move rapidly to make the project a fait accompli.[42] It established the China Yangtze Three Gorges Project Development Co. and began awarding a series of contracts for preliminary work. Construction began in December 1994, and many thousands of people have been resettled with miserly compensation.[43] The government tried to attract foreign support by making agreements with companies like Caterpillar, Ingersoll Rand, and General Electric of the United States, Atlas Copco of Sweden, and Krupp and Mannesmann of Germany. Some foreign groups refused, however, to participate.[44]

Li Peng, whose second term as prime minister ends in March 1998, wants to make the project unstoppable before he leaves office, and it appears that he will succeed. By early 1997, the project was beginning to look inevitable, and critics were shifting their attacks from the dam itself to the destruction of cultural artifacts by the reservoir. Ancient graves and cultural relics were already being destroyed by earth-moving equipment. In late 1996 and early 1997, a group of 56 prominent Chinese scholars protested to the Chinese government the destruction on a large scale of cultural relics in the Three Gorges area.[45] The chief engineer of the Three Gorges Project, Wei Tingcheng, angrily replied, "To tell the truth, the common people of China have such a low education level that they will not be able to enjoy these cultural relics. . . . Right now, to spend so much money to build museums does not conform to the interests of the state."[46]

The Chinese government continues to try to suppress all domestic criticism of the dam. The government has even tried to silence news about previous dam catastrophes in order to avoid speculation about the effects of a possible collapse of the Three Gorges Dam. In February 1995, Human Rights Watch/Asia issued a report accusing China of suppressing information about the collapse of two dams in 1975 that resulted in between 100,000 and 200,000 deaths.[47] A spokesman for the Chinese government finally admitted that the Banqiao and Shimantan dams on the Huai River in Henan Province had collapsed in 1975, but refused to give casualty figures, and maintained that a similar disaster could not happen at Three Gorges. Meanwhile, Human Rights Watch/Asia announced that "at least 179 people were detained in 1992 for opposing the dam."[48] In the areas where people are being displaced, the government has increased security to combat what it calls "vandalism and terrorism." One suspects that some of these vandals and terrorists were merely people trying to defend their homes, like the Mennonite farmers at Dneprostroi in the late 1920's, or the Tennessee Valley farmers in the early 1920's. Robin Munro, Hong Kong director of Human Rights Watch/Asia observed, "I can easily see the entire region becoming a police state within a police state." In February 1995, Human Rights Watch/Asia released a report in which it said the Chinese government was using forced labor in the dam's construction.[49]

Construction of the Three Gorges Dam is proceeding, but the project is so large and will take so many years (the estimated completion date is 2009) to complete that opportunities still exist for its cancellation. With the death of Deng Xiaoping in 1997, everyone knows that a leadership turnover in China is at hand, with an older generation being supplanted by a younger one, and critics of the dam hope that a change in policy toward such megaprojects may come with the leadership change. Unfortunately, there are even more narrowly trained engineers among the younger candidates for the Politburo.

Now I would like to shift to the Central Artery/Third Harbor Tunnel Project. It, too, is a megaproject, the largest public works project being pursued in the United States today. Although not as large as the Three Gorges Dam, at an estimated $10 billion, its expense is more than three-quarters of the officially stated price of that dam. One can be fairly confident that both estimates will be revised upwards.

One of the interesting aspects of the Central Artery/Third Harbor Tunnel Project is that it is a highway project being pursued in the heart of a city where only a few decades ago, opposition to highway building reached an apogee. A careful follower of the project observed in 1992, "one of the great ironies that I saw in my political lifetime was that we went from this almost hysterical opposition to the Third Harbor Tunnel in the mid-1970s to . . . [its] metamorphosis . . . [into] one of the great centerpieces of transportation policy." "Why was it so bad then and . . . it is not only a good thing now, it also is a great and necessary thing to have?"[50]

The answers to these questions mark a great transition that has occurred in American society toward megaprojects, one that can be characterized as a turn away from the imposition of such projects on an unwilling society to an inclusion of that society in the decision to build such projects and a thorough consultation with the society on the ways in which the construction will proceed. No longer are technical experts or political leaders given the right to make these decisions; the society must participate in the decisions as well.

One can not understand the present artery/tunnel project in Boston unless one sees that in the history of American public works, it is a "third-phase" project. The first phase embraced American history in the nineteenth century and the first half of the twentieth. This was a period in which Americans embraced technology uncritically and accepted the opinions of experts on how megaprojects should be built. The present elevated artery in Boston was built in the 1950's, when this faith was still strong. As Thomas P. "Tip" O'Neill Jr., speaker of the Massachusetts House of Representatives at the time, said "Everything was done by the greatest architects and engineers in the country."[51] In retrospect, however, the building of the central artery through the heart of downtown Boston in the 1950's was the beginning of the end of this era. The elevated roadway required significant "takings," the development jargon for forced eviction of local families. About 1,000 residential and commercial structures were taken by eminent domain, evoking an outcry among the affected local citizens, many of whom were Italians by ethnic origin. The artery was built nonetheless, but it turned out to be a disappointment. First of all, it blighted downtown Boston, dividing the central business district from the North End. Second, it did not work very well and was soon jammed with slow-moving traffic.

The solution to the traffic problem, the experts thought in the 1950's, was more highways—an Inner Belt that would create a ring road around the downtown core. This inner belt, connecting with highways 93 and 95 of the interstate system, would have required even more takings. In fact, it would have involved the demolition of about 3,800 homes in Roxbury, a predominantly black neighborhood; the Back Bay Fens, a park near the city's most prestigious cultural institutions; a sliver of affluent and vocal Brookline; and in working-class areas of Cambridge, Somerville, and Charlestown.[52]

To cut a very long and complicated story short, the plan for the Inner Belt provoked an outcry of opposition in the late 1960's and the formation of a political bloc that came to be known as "The Anti-Highway Coalition." Among the leaders of the coalition were Michael Dukakis, a young state representative from Brookline, and Fred Salvucci, an MIT-trained civil engineer whose grandmother's house was taken for a turnpike extension.[53] These men decided that the old way of building public works projects in this country must be abandoned. The Anti-Highway coalition, soon joined by now-congressman Tip O'Neill and senators Edward Kennedy and Edward Brooke, was successful. The coalition was victorious, and in November 1972, Boston authorities announced that no new highways would be built inside Route 128 around Boston.[54]

And yet within less than twenty years, groundbreaking began on the Central Artery/Third Harbor Tunnel Project, a megaproject of enormous dimensions that would greatly increase highway transit through central Boston. What happened?

Many things happened, but perhaps the most important was for experts to start paying attention to local citizens and to avoid the coerced taking of residences and businesses. The paradigm shift that occurred cannot best be described as attempting to mitigate the disruption that comes when public works projects are built. The shift is better described as the belief that engineering designs are actually improved by public participation. As Kenneth Kruckemeyer of the Department of Civil and Environmental Engineering at MIT recently stated: "Even a thoughtful engineer will not come up with a good design unless he works under constraints, of which public participation is one. . . . It's the learning process back-and-forth that allows a project to become better. . . . The only federal highway projects which have won awards for good design are those which have been sued in ways forcing changes in design."[55]

This philosophy, which is even yet not fully accepted in the United States, represents a dramatic shift from the doctrine of the early twentieth century that experts know best. Some experts are now saying publicly that they cannot do a good job unless they have public input. They need to know what the public wants and what it does not, and they can learn that only by having meaningful public participation—what Kruckemeyer calls "face-to-face contact and give-and-take." "The ultimate success of a project," he continues, "is how well it works for people."[56] One is reminded of Petr Pal'chinskii's unsuccessful call at the beginning of Soviet industrialization to build a society where industry served the need of the citizenry instead of the reverse.

In the case of the Central Artery/Third Harbor Tunnel Project in Boston, public hearings, community participation, and environmental impact statements were integral parts of the design project. By shifting the location of the Third Harbor Tunnel so that it emerges at Logan Airport instead of in a residential district, a way was found to build the tunnel that did not require the displacement of a single private residence. Opposition did, of course, exist, but the new principles of planning were based on trying, whenever possible, to meet the objections of the protestors. Fred Salvucci, who had been a well-known critic of highways in the 1960's, became Massachusetts's secretary of transportation. It was Salvucci's job to see if a new tunnel and central artery could be built with the support of the local community. His announced policy was "to bring the critics under the tent" by involving them in finding designs that were mutually acceptable.

Notice the difference between this attitude and that of engineers in the earlier part of the century. When the American engineer Frederick Winslow Taylor introduced what he called "scientific management" in the early decades of this century, he was convinced that only highly qualified specialists could provide useful input to technical decisions. Taylor did occasionally speak of the need for "cooperation" from workers, but he said that it meant for them "to do what they are told to do promptly and without asking questions or making suggestions."[57] Taylor believed that what he recommended was good for the workers, but he was the one who was to make all the decisions about what was good for them. This mind-set was adopted by engineers and government planners in the Soviet

Union, and it lives on today in China. When Wei Tingcheng, chief engineer of the Three Gorges Project in China, was asked not long ago if the project might not have harmful effects on the environment and on society, he replied: "Everything in the world has both positive and negative factors or influences, not to mention the Three Gorges Project, which is such a big project. Our policy decision is made on a scientific basis, which reflects the fundamental interests of the people."[58] Science, defined by specialists, was to be the determining element.

In conclusion, I would observe that the two greatest megaprojects currently being built in their respective countries—the Three Gorges Project in China and the Central Artery/Third Harbor Tunnel Project in the United States—are being pursued in radically different ways. The first is a surviving vestige of Soviet-style central planning by specialists who disdain the opinions of affected citizens. The second is emblematic of a new participatory planning process that assumes that obtaining public assent is necessary both for political and technical reasons; technical specialists are employees of the public, not their masters. Will the future move further in the direction of the second model, and not the first? In answer, I would guess that the Soviet/Chinese model is, indeed, a vestige, and that in all countries in the future, more public participation in the planning of megaprojects will occur. However, I think that it is possible that *both* the Three Gorges Project and the Third Harbor Tunnel represent outermost points of development that cannot be sustained in the future, although in very different, even opposite, directions. Three Gorges is probably the most extreme example of specialist-dominated central planning of the Soviet style that the world will witness in the foreseeable future. Resistance to it is growing even in China, under conditions of severe repression. Oddly enough, the Third Harbor Tunnel in Boston may also represent an outermost development. How realistic is it to imagine that future megaprojects of that size can be built without the taking of a single private residence? The special geographic conditions of the Boston project, especially the relative positions of Logan Airport and the harbor, permitted redesigns that minimized impact on the city's residents. (Other current megaprojects, such as the Sardar Sarovar Project, or the Narmada Dam, in India may fall on the spectrum between the opposite poles, in terms of public participation, of Three Gorges and the Third Harbor Tunnel.) There are other reasons why the Third Harbor Tunnel may be special. The political history of the project is quite unusual; its present phase is grounded in the power gained in the 1960's and 1970's by Democratic politicians like Dukakis, O'Neill, and Kennedy. A highway project in a new style was given birth by people who had revolted against a highway project in the old style. That particular sequence of events is not likely to be repeated. Furthermore, the politics of the Democratic Party, in which the Third Harbor Project is rooted, is less popular today than it was in the 1960's. Already there are calls by some members of the U.S. Congress to diminish federal support for the Boston project. Budget cutters point to the frequent cost overruns of the Third Harbor Tunnel Project. Doing things the new way is very expensive. Critics of environmental regulation can point out that the environmental impact statement for the Third Harbor Tunnel Project is 25 volumes long and cost a great deal of money. Should we spend that kind of money on a state that usually votes Democratic, and that was the only state in the Union to favor Dukakis in the 1988 presidential campaign?

Furthermore, cynics will say that the new "participatory

planning" has produced a new type of engineer, one who is as skilled at manipulating the public with public relations as the old type used to be at manipulating it by claiming superior technical knowledge. One engineer, Richard Meehan, published a book in 1981 entitled *Getting Sued, and Other Tales of the Engineering Life*, in which he observed:

> If you watch the lawyers, in time you learn the trick of trafficking in words. My firm has tripled in size (since we got sued). We write environmental impact reports now, some of them so big it takes more than one strong man to lift them, and we earn a good profit performing studies and analyses required (but I suspect not read) by bureaucrats. "Forensic engineering" I call it. It's the new way, and business is booming.[59]

At MIT, a special course was recently offered entitled "Managing Public Participation in Public Works Projects."[60] The word *management* here still has a Taylorist ring to it. In the old days, engineers talked of the "scientific management" of recalcitrant workers who tended to loiter, or "soldier," on the shop floor. Today engineers speak of managing the public, which tends to get obstreperous about large construction projects. This new type of engineer has learned the language of public hearings, of the courts, and of community relations.

But whatever the future of public works projects in America, the model of social engineering directed solely by experts that was taken to its extreme form in the Soviet Union and later in China is, we may hope, dead. Megaprojects are not merely technical and should not be planned and executed by technical specialists alone; they are highly political. The progressivists of the United States of the 1920's who thought that politics should be kept out of technology were wrong. The politics, if we mean local politics, must be put back in. The

Soviet Union claimed in its last decades that it was building "scientific communism," a society run on scientific principles and directed by specialists. The fate of that society emphasizes the point made by the chastened urban planners of MIT who now say "the engineer works best under constraints, and one of the most valuable of these constraints is public participation."[61]

Conclusions

In this book I have tried to answer the question "What have we learned about science and technology from the Russian experience?" After finishing the book my guess is that some American readers will say, "I have learned a lot more about Russia than I have about science and technology. The Russian experience is so unusual that it does not tell us much about science and technology." Such a conclusion would be, in my opinion, a serious mistake. It is based on two erroneous assumptions: (1) that science and technology can be separated from their societal context; and (2) that "real" science and technology can be found only in North America and Western Europe.

Let us for a moment imagine a scholar in Russia who has spent his or her professional life studying science and technology in the United States. Such a scholar would be my mirror image, since I am an American who has spent his professional life studying science in Russia. Let us assume that this Russian scholar published a book entitled "What have we learned about science and technology from the American experience?" In this book the Russian scholar describes a great many features of American science and technology, such as the entrepreneurism of their most successful practitioners, their competitiveness and the importance of proposal writing, their support by a mixture of state and private universities and foundations, the political lobbying for money and the ensuing struggles over such "big science" projects as the unsuccessful Superconductor Supercollider, the enormous importance of corporations and the market in influencing industrial research, the role of peer review in the awarding of grants and the promotion of researchers, and the involvement of science with religious and ethical disputes such as those over creationism, Darwinism, genetic engineering, abortion, the human genome project, fraud in research, and animal rights. One can understand a Russian reader who, after reading this book, might say, "I learned a lot more about the United States than I did about science and technology."

The fact that American and Russian readers might draw such similar conclusions pointing in opposite directions illustrates the inextricable linking of science and society that characterizes the end of the twentieth century. In my opinion, the study of these linkages is an exceptionally fruitful endeavor because it informs us about *both* science and society. The Russian example is valuable precisely because the social and economic factors in Russia were strikingly different from those in the United States and the West, and the effects of these different factors are therefore particularly visible. The result of the analysis of science in Russia is not an emphasis on the uniqueness of science there, but a clarification, by contrast, of the importance of social and economic factors on the science of *all* countries.

I would now like to turn to the questions I ask in each of my chapter titles:

Is science a social construction? The experience of Russian science provides two different answers to this question for two different groups of people. The different answers are required by the assumptions that these people are likely to possess before they consider the question. For natural scientists and those members of the educated public who often believe that science is "truth," that it is an objective description of nature, the Russian experience demonstrates that science contains many elements of social, political, philosophical, and ideological influence. Even the best Russian science, including internationally recognized schools of thought in "hard" sciences like physics and mathematics, demonstrates the unmistakable influence of social factors.

For extreme social constructivists, those people who think that science is a social product to the same degree as literature or philosophy, the Russian experience provides a different answer. It demonstrates that sometimes social influences lead so far away from empirical evidence and cognitive rigor that "reality" eventually demands a correction. That happened in Soviet biology, and it also happened in a number of other fields. The victory, after a 30-year struggle, of the critics of Lysenkoism in the Soviet Union over its supporters—who had enormous social and political advantages—is an instance where empirical evidence and cognitive rigor vanquished a rival theory constructed largely for social reasons.

The experience of Russian science provides useful insights into the great debate over the social construction of science that has occurred recently in the United States and Europe. Science is a social construction in the sense that scientists, members of society, make it, and they are inevitably influenced by social factors in the process. Contrary to the views of many natural scientists, the influence of social factors extends to the core of science itself, the theories of explanation around which the scientists of a field frequently unite. But natural science deals with objective reality to a much higher degree than the humanities, and therefore empirical evidence is often much more influential. The leash that ties scientific theories to reality is far longer and slacker than most people know, but it does exist.

The history of Russian science contains many paradoxes, and none is perhaps greater than the fact that Soviet scientists, working in an atmosphere of political oppression, produced a philosophy of science, dialectical materialism, that comes as close as any to capturing the essence of modern science. Dialectical materialism draws heavily on the achievements of modern science. This philosophy of science is discussed both in this book and in my other publications.[1] The full appreciation of the significance of this philosophy of science is only beginning in the Western world. The eminent evolutionary biologist Ernst Mayr is one of the few non-Marxist Western scientists who have noticed the importance of dialectical materialism. Mayr speaks of its similarity to his own philosophical view, which he calls "organicism."[2] Organicism, however, is a philosophy of *biology* rather than a general philosophy of nature. Dialectical materialism has a larger reach than organicism, including both the biological and the physical sciences.

Are science and technology westernizing influences? In the case of the Soviet Union, science and technology helped to reduce the revolutionary exceptionalism that its leaders originally claimed. Science and technology were, by their nature, more closely linked to developments elsewhere than most other fields of intellectual endeavor, and scientists and engineers (especially scientists) often emerged as dissidents and critics in Soviet society. The majority of the members of the scientific and technical community in the Soviet Union yearned for a more democratic and free political system, and many of them helped to bring such a system about. But one should be careful about concluding that scientists and engineers will always side with Western concepts of democracy and freedom. Especially when their interests are threatened, scientists and engineers, like any other professional group, may be tempted to support a nondemocratic political leader or party promising to rescue them and give strong financial assistance to research and development. That danger currently exists in Russia.

How robust is science under stress? The experience of Russia demonstrates that science is incredibly robust. Far from being a delicate flower, the most fragile product of ultimate civilization that many of its defenders like to describe, science has become an integral part of modern society, perhaps its most essential part, and it will not disappear as long as modernity continues to exist.

The knowledge industry in modern societies is no longer a minor affair run by an intellectual elite, an activity that might be considered by pragmatic leaders as expendable; it is a mammoth enterprise on a par with heavy industry, and just as necessary to the country in which it is situated. By the late 1980's, there were more employees in "science and science services" in the Soviet Union than there were in the fuel, energy, and metallurgical industries combined.[3] But not just the Soviet Union possessed an enormous knowledge industry. In the United States, the TIAA/CREF (Teachers Insurance and Annuity Association/College Retirement Equities Fund), a teachers' and researchers' organization, had become by about the same time the single largest private pension fund in the country. (Historians, unfortunately, have not caught up with the importance of the knowledge industry to modern society; they continue to emphasize political history and labor history at a time when the history of science, technology, and education should be growing and ever more influential fields.)

Today, every industrial society needs science and will find a way to support it, even if that same society frequently abuses science. The Soviet Union politically repressed science atrociously while simultaneously supporting it financially more fulsomely, relative to its resources, than any other country in history. The sobering conclusion that we must draw, in terms of scientific results, is that the support counted for more than the repression. Under the Stalinist system, the Soviet Union became a major scientific nation, developing nuclear weapons and other sophisticated armaments and sending the first artificial satellites into orbit. And its achievements were not only in strategic areas. In abstract and fundamental fields, such as mathematics and theoretical physics, the Soviet Union became a world leader.

An image that lingers in the mind is that of the young scientist Andrei Sakharov gazing out the window at rows of political prisoners being marched to work at his scientific installation, while, at the same time, Sakharov was sitting at his desk developing the Tokamak model of controlled nuclear fusion, a model that dominated research in the field in all countries for over a generation and still does today. No more striking combination of oppression and scientific creativity can be imagined.

The robustness of science, its ability to survive and even prosper in the most inhospitable environments, leads us eventually to a clearer justification for democracy and human rights. Westerners have often in the past maintained that freedom and human rights are necessary for a flourishing and creative economy. This message is now being propagated by the United States throughout the world, with considerable success. The Soviet experience shows that there is, indeed, a positive correlation between freedom, on the one hand, and a prosperous economy and a creative culture, on the other. Soviet culture and science were not as productive as they should have been, considering the enormous resources they consumed; the lack of freedom and an open market were certainly important causes of this failure. But Soviet science was sufficiently successful that we can see the dangers of the utilitarian argument, the belief that democracy and human rights are justified by the fruits they produce. It is not inconceivable that some day a society will appear (China in the twenty-first century?) that is quite successful in science and technology but neither democratic nor observant of human rights. The Soviet Union came close enough to that model to point to the disturbing possibilities. Should that happen, if we have invested too heavily in the utilitarian argument, we would be in a vulnerable position.

Democracy and human rights are more important than science and should take priority over it. The primary justification for democracy and human rights is ethical, not utilitarian. Westerners sometimes have difficulty seeing this priority of values because they have been blessed with situations in which the positive correlation of science and freedom was obvious. One of the lessons of the Soviet Union is that this correlation is not as direct and obvious as most of us thought, and that it is possible that some day we might have to decide which we value most, science or democracy. By thinking through this complicated relationship, it may now be hoped that we shall be able better to see that democracy and human rights are considerably more important than science. We should continue to try to have them all—democracy, human rights, and science—but we should know which comes first.

If one myth that the Russian experience seriously questions is that science cannot exist without freedom; another myth that it undermines is the view of the "connectivity" of the various fields of knowledge. It is often said that if one field of knowledge is destroyed or repressed, the effects will spread to all other fields. Academic freedom is thus seen as a connected whole, all of the parts depending on the other. The Russian experience points to a more selective influence of political repression. The humanities and social sciences suffered much more in the Soviet Union than most of the natural sciences. In the 1970's, when the National Academy of Sciences of the United States sponsored a study that concluded that in certain areas of mathematics and physics, the Soviet Union was a "world leader," no one would have said the same thing about Soviet history or sociology. The humanities and social sciences in the Soviet Union at this time were largely a wasteland. Genetics had also been repressed until a few years earlier. But while several fields were nearly destroyed in the Soviet Union, Soviet physics and mathematics were rising to prominence and attracting attention throughout the world. A realistic view of the history of science simply has to deal with these uncomfortable facts, because they cannot be dodged.

Should we conclude, then, that science is so robust that we need not worry about coddling it, giving it intellectual freedom and financial support? The Russian experience points to a strong distinction between those conditions that are necessary for the survival, even prospering, of science, and those that are necessary for its most creative achievements. The enormous Soviet scientific establishment, the world's largest, performed rather well in many areas, provided for the nation's military strength, and supplied most of the needs of heavy industry. But it did not do well in terms of intellectual breakthroughs or outstanding achievements. No matter what criterion of evaluation one chooses—Nobel prizes, citation indices, international patent licenses, or membership in international honorary societies—the conclusion is inescapable, as discussed in chapter 4, that the Soviet Union was underachieving, not receiving as much outstanding science and technology as it should have in return for its enormous expenditures on research and development. Political freedom may not be as necessary for the development of natural science as many of its advocates have claimed, but a combination of political freedom and generous financial support *are* necessary for the most creative achievements. One of the tragedies of Russian history is that science there has never enjoyed both financial support and political freedom, either under the Soviet system or today, although, in chronological sequence, it had first the one and then the other. What is remarkable about Russian science is how much it has accomplished despite this problem.

How willing are scientists to reform their own institutions? The experience of Russia points to the conclusion that scientists are no more willing to reform their own institutions than any other interest group, and more articulate in their defense than most. The system of fundamental research that developed in the Soviet Union was in deep need of reform, but the scientists who controlled it have successfully resisted major changes even after the collapse of the system that created it. In this sense, they have behaved conservatively, like privileged groups throughout history. However, scientists defend some privileges much more stoutly than others. Most important of all is their sense of status. Labor unions often fight for wage increases and material benefits. Scientists rarely do so. The scientists in the Soviet Union who controlled the Academy of Sciences lost most of their financial advantages after the country collapsed, but they fought most strenuously to retain their nonmonetary perquisites and influences, especially their roles as the administrators and leaders of the science establishment. As a result,

Russia is today the only major country in the world in which several hundred leading, and often quite senior, scientists, chosen by themselves, are in charge of the fundamental science establishment, directing its laboratories and institutes. They fiercely defend that privilege even at a time when they cannot pay the researchers who work in those laboratories and institutes.

Who should control technology? The example of the Soviet Union shows that technology, especially the megaprojects that reshape the lives of entire communities, is far too important to leave to the technical specialists. The leaders of the Soviet Union promised to construct a society that would be governed scientifically, and in which the major decisions about modernization and industrialization would be made by qualified experts. The disastrous results of that effort help us to understand that when the welfare of entire communities and nations are at stake, the members of those communities and nations, including those who are not technically trained, must be brought into the decision process about questions of technology. A number of countries in the world today have yet to learn this lesson, and even in the United States, the consensus on this issue is insecure.

The Soviet Union is gone, and most people who recall it today do so only with the thought that it was a totalitarian horror, best forgotten. The Soviet Union has a far greater significance. It was, during its existence, the only advanced industrial society based on principles—economic, social, philosophical—strikingly different from those of other industrial nations. That fact presents us, from an anthropological and historical perspective, with rich analytical possibilities. If we examine those aspects of Soviet culture that performed best and that we also vigorously promote—and science is probably the most prominent of these—we can gain much from asking how much and in what ways those aspects of Soviet culture were different from ours because of the unique society in which they were situated. If we answer this question, we shall learn as much about ourselves as we shall about the Soviet Union. We can understand how much *our* science and *our* culture are conditioned by *our* society only if we have another society and another culture with which we can appropriately draw comparisons. The Soviet Union may well have been one of the last modern alternatives to dominant Western patterns with which such comparisons of science can be made. In that sense, the study of Soviet science is also a study of *our* science.

Notes

Notes

Preface

1. For a study of the growth of Soviet research personnel, see Louvan E. Nolting and Murray Feshbach, "R and D Employment in the USSR," *Science* (Feb. 1, 1980): 493–503.

Chapter 1

1. Although Robert Merton himself, in his earliest work, did. This effort to analyze the content of science was abandoned by most of his followers. See Robert King Merton, *Science, Technology and Society in Seventeenth-Century England* (1938; New York: H. Fertig, 1970).

2. R. Whitley, "Black Boxism and the Sociology of Science: A Discussion of the Major Developments in the Field," in *The Sociology of Science*, ed. P. Halmos, *Sociological Review* monograph no. 18 (Keele, Eng.: University of Keele, 1972), pp. 61–92.

3. Peter L. Berger and Thomas Luckmann, *The Social Construction of Reality: A Treatise in the Sociology of Knowledge* (Garden City, N.Y.: Doubleday, 1966). In the discussion here, I am following the evolution of the field traced in Trevor J. Pinch, "Recent Sociology of Science: A Review" (paper presented at the workshop "New Approaches to the History and Social Study of Science and Technology," St. Petersburg, Russia, June 1994).

4. Bruno Latour and Steven Woolgar, *Laboratory Life: The Construction of Scientific Facts* (Princeton: Princeton University Press, 1986); Steven Shapin, *Leviathan and the Air Pump: Hobbes, Boyle and the Experimental Life* (Princeton: Princeton University Press, 1985);

Andrew Pickering, *Constructing Quarks: A Sociological History of Particle Physics* (Chicago: University of Chicago Press, 1984); Peter Galison, *How Experiments End* (Chicago: University of Chicago Press, 1987); and Mario Biagioli, *Galileo Courtier: The Practice of Science in the Culture of Absolutism* (Chicago: University of Chicago Press, 1993) have been particularly influential.

5. Alan D. Sokal, "Transgressing the Boundaries—Toward a Transformative Hermeneutics of Quantum Gravity," *Social Text* (Spring/Summer 1996): 219–52. Sokal revealed that this article was a hoax designed to parody science studies in his "A Physicist Experiments with Cultural Studies," *Lingua Franca* (May/June, 1996): 62–64.

6. Pinch, "Recent Sociology of Science," p. 6.

7. See the analysis in Susan Solomon, "Reflections on Western Studies of Soviet Science," in *The Social Context of Soviet Science*, ed. Linda Lubrano and Susan Solomon (Boulder, Colo.: Westview Press, 1980), pp. 1–29.

8. Valentin F. Turchin, "The Exclusion Principle," *Nature* 331 (Jan. 7, 1988): 23–24.

9. See especially Loren R. Graham, *Science and Philosophy in the Soviet Union* (New York: Knopf, 1972); and *Science, Philosophy, and Human Behavior in the Soviet Union* (New York: Columbia University Press, 1987).

10. One of the major Western interpreters of Vygotsky has been Michael Cole, who studied as a postdoctoral fellow in Moscow under Alexander Luria. Luria persuaded Cole to publish a book by Vygotsky in the West. Together with Vera John-Steiner, Sylvia Scribner, and Ellen Souberman, Cole coedited a volume of Vygotsky's essays, *Mind in Society* (Cambridge, Mass.: Harvard University Press, 1978). Cole has written extensively on Vygotsky elsewhere, most recently in his *Cultural Psychology: A Once and Future Discipline* (Cambridge, Mass.: Harvard University Press, 1996).

11. Quoted in James V. Wertsch, "L. S. Vygotsky's 'New' Theory of Mind," *American Scholar* 57 (Winter 1988): 81.

12. Jerome S. Bruner, "Introduction," in L. S. Vygotsky, *Thought and Language*, ed. and trans. Eugenia Hanfmann and Gertrude Vakar (Cambridge, Mass: MIT Press, 1962), pp. vi, x.

13. Wertsch, "L. S. Vygotsky's 'New' Theory of Mind," p. 87.

14. Here I am following the treatment in my *Science in Russia and*

the Soviet Union: A Short History (Cambridge: Cambridge University Press, 1993), pp. 103–8.

15. The translators commented: "Although our more compact rendition would be called an abridged version of the original, we feel that the condensation has increased clarity and readability without any loss of thought content or factual information" (Eugenia Hanfmann and Gertrude Vakar, translators' preface to Vygotsky, *Thought and Language*, p. xii).

16. Vygotsky, *Thought and Language*, pp. 49–51.

17. See Katerina Clark and Michael Holquist, *Mikhail Bakhtin* (Cambridge, Mass.: Harvard University Press, 1984), esp. pp. 229–30.

18. Stephen W. Hawking, *A Brief History of Time: From the Big Bang to Black Holes* (New York: Bantam Books, 1988), pp. 48–50, 104–5, 130–32. For a discussion of the connection between the big bang theory and Western cultural and religious attitudes, see Daniel Kevles, "The Final Secret of the Universe?" *New York Review of Books*, May 16, 1991, pp. 27–32.

19. In a book entitled *Psychology and Marxism*, Luria called for "a radical reworking of psychology in terms of the scientific method of dialectical materialism." See Luria in *Psikhologiia i marksizm*, ed. K. P. Kornilov (Leningrad, 1925), pp. 476–80. Luria became one of the great psychologists of the Soviet Union and has been the subject of several studies by the American psychologist Michael Cole. See *The Selected Writings of A. R. Luria*, ed. Michael Cole (White Plains, N.Y.: Sharpe, 1978); Michael Cole, "Introduction," in A. R. Luria, *Cognitive Development* (Cambridge, Mass.: Harvard University Press, 1976); and id., "Afterword," in A. R. Luria, *The Making of Mind* (Cambridge, Mass.: Harvard University Press, 1979). And see also Graham, *Science, Philosophy, and Human Behavior in the Soviet Union*, pp. 184–91.

20. Rubinshtein was the founder of the department of psychology at Moscow University and his works have been published in many languages; he is still recognized as one of the most influential voices in psychology in Russia. He commented that his interpretation was heavily influenced by the "dialectical materialist understanding of the determination of psychic phenomena." See S. L. Rubinshtein, *Printsipy i puti razvitiia psikhologii* (Moscow, 1959), p. 3. For a discussion

of the relationship of his psychological views and Marxism, see Graham, *Science, Philosophy, and Human Behavior in the Soviet Union,* pp. 176–84.

21. Leont'ev was best known for his "activity" psychology, which he closely connected to Marxism. See his *Activity, Consciousness, and Personality,* trans. Marie J. Hall (Englewood Cliffs, N.J.: Prentice-Hall, 1978). A discussion of the influence of Marxism in his works is to be found in Graham, *Science, Philosophy, and Human Behavior in the Soviet Union* , pp. 211–14.

22. Anokhin was well known outside the Soviet Union as an extender of the Pavlovian tradition, which he tied closely to Marxism. See his praise of dialectical materialism in "Metodologicheskii analiz uzlovykh problem uslovnogo refleksa," in *Filosofskie voprosy fiziologii vysshei nervnoi deiatel' nosti i psikhologii,* ed. P. N. Fedoseev et al. (Moscow, 1959), p. 158; and in Graham, *Science, Philosophy, and Human Behavior in the Soviet Union,* pp. 200–211.

23. Serebrovskii was an outstanding geneticist and a sincere Marxist. He first formulated the concept of the "gene pool" (*genofond*), which was brought from the Soviet Union to the United States by Theodosius Dobzhansky, who translated it into English. For his views that classical genetics and Marxism were mutually reinforcing, see his "Teoriia nasledstvennosti Morgana i Mendelia i marksisty," *Pod znamenem marksizma,* 1926, no. 3.

24. Dubinin, one of the most prominent geneticists in the Soviet Union in the post-Lysenko period, wrote many articles and books connecting Marxism to his science. See, e.g., his "Sovremennaia genetika v svete marksistskoleninskoi filosofii," in *Lenin i sovremennoe estestvoznanie,* ed. M. E. Omel'ianovskii (Moscow, 1969), pp. 287–311.

25. Aleksandrov was the founder of the Soviet school of geometry in the large and has published many articles on the relationship of Marxism to mathematics and physics. See, e.g., his "Dialektika i nauka," *Vestnik akademii nauk SSSR,* 1957, no. 6: 3–17.

26. Kolmogorov, a leader in mathematics in this century, wrote about Marxism and mathematics in several different places, including *Bol'shaia sovetskaia entsiklopediia,* vol. 26 (Moscow, 1954), p. 464; ibid., vol. 15 (Moscow, 1974), p. 467; and "Matematika stalinskoi epokhi," Arkhiv MGU, fond 2, op. 4, ed. khr. 3, esp. pp. 11–12.

27. Shmidt, a mathematician and astronomer, was a very famous

man in early Soviet history. His private papers show that he was a serious student of Marxism and science, as do many of his publications. See Archives of the Academy of Sciences of the USSR, fond 496, op. 1, ed. 212; and his *A Theory of the Earth's Origin* (Moscow, 1958), pp. 83–84.

28. Semkovskii was the first Soviet physicist to give an analysis of Einstein's relativity physics, and he connected it to Marxist materialism. See his *Dialekticheskii materializm i printsip otnositel'nosti* (Moscow and Leningrad, 1926).

29. Blokhintsev, director of the Joint Institute of Nuclear Research at Dubna, went to great lengths to develop an interpretation of quantum mechanics that he believed was compatible with Marxism. See his *The Philosophy of Quantum Mechanics* (Dordrecht, Holland, and New York: Humanities Press, 1968).

30. Naan was a prominent physicist in the Soviet Union who often connected his field to Marxism. See, e.g., his "O beskonechnosti vselennoi," *Voprosy filosofii*, 1961, no. 6: 93–105.

31. Ambartsumian, former president of the International Astronomical Union, was a particularly outspoken proponent of the benefit of Marxism to astronomy. See, e.g., his "Nekotorye voprosy kosmogonicheskoi nauki," *Kommunist*, 1959, no. 8: 86–96.

32. Zel'manov, a well-known scientist at the Shternberg Astronomical Institute of Moscow University, often spoke and wrote on what he saw as the mutually beneficial relationship of Marxism and science. See, e.g., his "Mnogoobrazie material'nogo mira i problema beskonechnosti Vselennoi," in *Beskonechnost i Vselennaia* (Moscow, 1969), pp. 279–324.

33. This definition of mathematics occurs in all three editions of Kolmogorov's article. See *Bol'shaia Sovetskaia Entsiklopediia* (hereafter *BSE*), vol. 38 (Moscow, 1938), col. 359; *BSE*, vol. 26 (Moscow, 1954), p. 464; *BSE*, vol. 15 (Moscow, 1974), p. 467.

34. This statement was added to the last two editions: *BSE*, vol. 26 (Moscow, 1954), p. 464; *BSE*, vol. 15 (Moscow, 1974), p. 467.

35. F. P. Ramsey, "Mathematics, Foundations of," *Encyclopedia Britannica*, vol. 15 (1941), p. 83.

36. Ibid.

37. A. N. Whitehead, "Mathematics, Nature of," *Encyclopedia Britannica*, vol. 15 (1941), pp. 87–88.

38. Stephen Toulmin, "How Can We Reconnect the Sciences with the Foundations of Ethics?" in *Knowing and Valuing: The Search for Common Roots*, ed. H. Tristram Engelhardt Jr. and Daniel Callahan (Hastings-on-Hudson, N.Y.: Hastings Center, 1980), p. 47.

39. Ibid., p. 48.

40. Loren Graham, *Science in Russia and the Soviet Union: A Short History* (Cambridge: Cambridge University Press, 1993), p. 102.

41. G. B. Zhdanov, "Razmyshleniia o 'goriachikh tochkakh' estestvoznaniia i putiakh razvitiia dialekticheskogo materializma" (paper presented at conference on "Philosophy of Science of the XXth Century: Results and Prospects," Institute of Philosophy, Russian Academy of Sciences, Moscow, Oct. 10, 1996), pp. 1–2.

42. A. D. Aleksandrov, an outstanding mathematician and former rector of Leningrad University, continues to have an interest in dialectical materialism, as do a number of other scientists (personal communications).

43. See Graham, *Science and Philosophy in the Soviet Union*; *Science, Philosophy, and Human Behavior in the Soviet Union* ; *Science in Russia and the Soviet Union*, esp. pp. 99–120.

44. David Joravsky, *The Lysenko Affair* (Cambridge, Mass.: Harvard University Press, 1970); Zhores Medvedev, *The Rise and Fall of T. D. Lysenko* (New York: Columbia University Press, 1969); Graham, *Science, Philosophy, and Human Behavior in the Soviet Union*. Also see Valerii Soifer, *Vlast' i nauka: istoriia razgroma genetiki v SSSR* (Tenafly, N.J.: Hermitage, 1989), published in English as *Lysenko and the Tragedy of Soviet Science* (New Brunswick, N.J.: Rutgers University Press, 1994); Mark Adams, "Genetics and Molecular Biology in Khrushchev's Russia" (Ph. D. diss., Harvard University, 1973); Nikolai Krementsov, *Stalinist Science* (Princeton: Princeton University Press, 1997); and Dominique Lecourt, *Proletarian Science? The Case of Lysenko* (London: NLB, 1977).

45. Robert Conquest, *The Harvest of Sorrow: Soviet Collectivization and the Terror-Famine* (New York: Oxford University Press, 1987).

46. For Lysenko's characterization of the geneticists as "man-haters" and his disdain for *Drosophila*, see Medvedev, *Rise and Fall of T. D. Lysenko*, pp. 81, 125.

47. Laura Belin, "Secret Agents 301, 329, and 345: The Introduction of Hybrid Seed Corn in the USSR" (senior thesis, Harvard University, 1991).

48. Lily Kay's forthcoming study of the history of molecular biology points to the influence of information theory, some of it developed for wartime purposes, in its formation. Evelyn Fox Keller has emphasized the significance of gender issues in the history of biology and genetics. Robert Kohler highlights the role of patronage, especially by the Rockefeller Foundation, on the early history of molecular biology. Many other historians are finding other social influences on the history of modern biology. See Lily Kay, *Who Wrote the Book of Life? A History of the Genetic Code* (Chicago: University of Chicago Press, forthcoming); Evelyn Fox Keller, *A Feeling for the Organism: The Life and Work of Barbara McClintock* (San Francisco: W. H. Freeman, 1983); Keller, *Reflections on Gender and Science* (New Haven: Yale University Press, 1985); Robert E. Kohler, *Partners in Science: Foundations and Natural Scientists, 1900–1945* (Chicago: University of Chicago Press, 1991); Lily Kay, *The Molecular Vision of Life: Caltech, the Rockefeller Foundation, and the Rise of the New Biology* (New York: Oxford University Press, 1993); and Pnina Abir-Am, "'New' Trends in the History of Molecular Biology," in *Historical Studies in the Physical and Biological Sciences*, vol. 26, pt. 1 (1995), pp. 167–96.

49. L. Fleck, *Genesis and Development of a Scientific Fact* (Chicago: University of Chicago Press, 1979).

50. The metaphor of magnets was suggested to me by Peter Galison, to whom I am indebted for an early critique of this chapter (discussions with Peter Galison, spring 1997, and especially his e-mail message of Mar. 26, 1997).

51. David Bloor, *Knowledge and Social Imagery* (1976; 2d ed., Chicago : University of Chicago Press, 1991), p. 166.

52. Douglas Weiner, *A Little Corner of Freedom: Russian Nature Protection from Stalin to Gorbachev* (Berkeley: University of California Press, forthcoming).

53. Karl Hall, forthcoming doctoral diss., History of Science Department, Harvard University.

54. Esther R. Phillips, "Nicolai Nicolaivich Luzin and the Mos-

cow School of the Theory of Functions," *Historia Mathematica,* 5, no. 3 (August 1978): 275–305.

55. Bloor, *Knowledge and Social Imagery*, p. 166.

Chapter 2

1. Medvedev wrote that before 1945 as a young student he thought that the controversy over genetics was "a real scientific debate" but, studying the issues more closely, "I understood that the main aim of Lysenko and his followers was anything but elucidation of scientific truth." His doubts about Lysenko constantly grew and eventually led him to become a dissident in the Soviet Union, persecuted by the secret police. He graduated from studies of science to the study of Soviet politics, and became one of the most outspoken domestic critics of the Soviet system. Medvedev also describes how Sakharov began to oppose Lysenko and his supporters, such as N. Nuzhdin, in Academy elections in the early 1960's. These actions were among Sakharov's first acts of political protest, and, as in the case of Medvedev, helped push him to the side of political dissidence. See Zhores Medvedev, *The Rise and Fall of T. D. Lysenko* (New York: Columbia University Press, 1969), esp. pp. 106, 215–17.

2. An informative discussion of the impact of information technologies on Russian society can be found in S. Frederick Starr, "New Communications Technologies and Civil Society," in *Science and the Soviet Social Order*, ed. Loren R. Graham (Cambridge, Mass.: Harvard University Press, 1990), pp. 19–50. Also see Wilson P. Dizard and S. Blake Swensrud, *Gorbachev's Information Revolution: Controlling Glasnost in a New Electronic Era* (Boulder, Colo.: Westview Press, 1987); Seymour Goodman, "Information Technologies and the Citizen: Toward a 'Soviet-Style Information Society,'" in *Science and the Soviet Social Order*, ed. Graham, pp. 51–67; and Richard W. Judy, *Implications of the Information Revolution for Soviet Society* (Indianapolis: Hudson Institute, 1989).

3. Lenin often expressed his opinion that wars under capitalism are inevitable. See, e.g., V. I. Lenin, *Collected Works*, 2d rev. ed. (Moscow: Foreign Languages Publishing House, 1960), 8: 53; 13: 80; 21: 39. Less frequently cited is his belief that socialists should nonetheless try to avoid wars. See ibid., 21: 299. When Soviet Russia needed a

breathing space in 1918, Lenin supported the Brest-Litovsk peace against the advice of his colleagues Bukharin and Trotsky.

4. David J. Dallin, *Soviet Foreign Policy after Stalin* (Philadelphia: Lippincott, 1961), p. 323. John Lewis Gaddis has maintained that "Cold War" is a misnomer for the period after World War II and has emphasized the stability of great-power relations during the period. See his book *The Long Peace: Inquiries into the History of the Cold War* (Oxford: Oxford University Press, 1967).

5. Nikita Khrushchev, *For Victory in Peaceful Competition with Capitalism* (New York: Dutton, 1960), p. xv.

6. Personal conversations with Paul Doty; also, Bernd W. Kubbig, *Communicators in the Cold War: The Pugwash Conferences, the U.S.-Soviet Study Group and the ABM Treaty: Natural Scientists as Political Actors: Historical Successes and Lessons for the Future* (Frankfurt a/M: Peace Research Institute Frankfurt, 1996).

7. Gregory Crowe, Department of the History of Science, Harvard University, is writing a history of Soviet computing. I am grateful to him for assistance on this section.

8. Paul R. Josephson, *New Atlantis Revisited: Akademgorodok, the Siberian City of Science* (Princeton: Princeton University Press, 1996); Douglas Weiner, *A Little Corner of Freedom: Russian Nature Protection from Stalin to Gorbachev* (Berkeley: University of California Press, forthcoming); Marshall Goldman, *The Spoils of Progress: Environmental Pollution in the Soviet Union* (Cambridge, Mass.: MIT Press, 1972); Thane Gustafson, "Environmental Issues Rise to Official Legitimacy," in his *Reform in Soviet Politics: Lessons of Recent Policies on Land and Water* (Cambridge: Cambridge University Press, 1961), pp. 39–52; Philip R. Pryde, *Environmental Management in the Soviet Union* (Cambridge: Cambridge University Press, 1991).

9. Oleg Volkov, "Tuman nad Baikalom," *Literaturnaia gazeta,* no. 16 (Feb. 8, 1965): 1–2.

10. See, e.g., his "Lessons of Lake Baikal," *Literaturnaia gazeta,* no. 41 (Oct. 11, 1967): 12; *Current Digest of the Soviet Press* 19, no. 48: 7.

11. See the discussion in Gustafson, "Environmental Issues," esp. pp. 46–50.

12. Volkov, "Lessons of Lake Baikal."

13. A. Merkulov, "Alarm from Baikal," *Pravda,* Feb. 28, 1965, p. 2, translated in *Current Digest of the Soviet Press* 17, no. 9 (1965): 25.

14. See, e.g., Romuald J. Misiunas, "The Baltic Republics: Stagnation and Strivings for Sovereignty," in *The Nationalities Factor in Soviet Politics and Society*, ed. Lubomyr Hajda and Mark Beissinger (Boulder, Colo.: Westview Press, 1990), pp. 214–15.

15. "Akademik Abrikosov nakhodit pliusy v 'utechke mozgov' na zapad," and A. Abrikosov, "Ia nikogda ne vernus' v Rossiiu," *Izvestiia*, May 5, 1993, p. 5. The condensed version appeared in Sergei Leskov, "America's Soviet Scientists," *New York Times,* July 15, 1993, p. A25. See also replies by Roald Hoffmann and Alexander Migdal, *New York Times,* July 25, 1993, p. 16. A complete version in Russian, with replies by Iakov Al'pert, Aleksandr Migdal, and the editors was published in *Vremia i my*, 1993, no. 12: 150–62.

16. *Vremia i my*, 1993, no. 12: 154.

17. Ibid., pp. 154–55.

18. Alexander Migdal, "Russian Scientists Are an Endangered Species," *New York Times,* July 25, 1993, p. 16.

19. *Vremia i my*, 1993, no. 122: 157.

20. Ibid., p. 161.

21. S. Moskvitianin, "Tainaia diplomatiia dzhordzha sorosa," *Molodaia gvardiia*, 1994, no. 2: 154–58.

22. Ibid., p. 158.

23. "FSK obespokoena aktivnost'iu amerikanskikh issledovatelei v Rossii," *Nezavisimaia gazeta*, Jan. 10, 1995, p. 3.

24. Ibid.

25. Some prominent examples of articles about the "Soros affair" are Vera Romanova, "Duma o Sorose: glubokaia blagodarnost' za vklad v sokhranenie," *Segodnia*, Feb. 23, 1995; Andrei Vaganov, "Sorosa—poblagodarit', Stepashina—prizvat' k otvetu," *Nezavisimaia gazeta*, Feb. 22, 1995; Valentin Zhdanko, "Dzhorzh Soros vstretilsia s presidentom Belorussii: Amerikanskii finansist obidelsia na rossiiskie spetssluzhby i sobiraetsia zakryt' fond podderzhki nauki v Moskve," *Segodnia*, Jan. 1, 1995; Kim Smirnov, "Pochemu darenyi kon' nepremenno dolzhen byt' troianskim? Dzhorzh Soros—Santa Klaus dlia uchenykh byvshego SSSR," *Novaia ezhednevnaia gazeta,* Jan. 11, 1995; "Shpionskie strasti," *Poisk*, Jan. 14–20, 1995; "My -za otkrytoe obsuzhdenie," *Novaia ezhednevnaia gazeta*, Jan. 25, 1995; Mariia Smir-

nova, "Uchenye zhdat pomoshchi ne tol'ko ot Sorosa," *Kommersant*, Jan. 25, 1995; "Fond Sorosa prekrashchaet finansirovanie rossiiskoi nauki," and Vitalii Gol'danskii, "Pis'mo v Izvestiia," *Izvestiia*, Jan. 26, 1995; "Vokrug fonda Sorosa," *Izvestiia*, Jan. 27, 1995; "Skandal soedinaet kontinenty," *Poisk*, Jan. 28–Feb. 3, 1995; "Iskat' shpionov nado, no tam, gde oni est'," *Rossiiskie vesti* (Feb. 3, 1995); "Dz. Soros i ego sorosiata," *Sovetskaia Rossiia*, Feb. 11, 1995; Mikhail Berger, "FSK podozrevaet pravitel'stvo v finansirovanii shpionazha na territorii Rossii," *Izvestiia*, Jan. 18, 1995; Nikolai Vorontsov, "O 'vklade' chekistov v otechestvennuiu nauku," *Literaturnaia gazeta*, Jan. 25, 1995; "Deputaty reshili zashchitat' fond ot spetssluzhb," *Kommersant*, Feb. 21, 1995; Andrei Vaganov, "Uchenye protiv kontrrazvedki," *Nezavisimaia gazeta*, Feb. 1, 1995; "Pravitel'stvo blagodarit MNF, a FSK podozrevaet ego v shpionazhe," *Novaia ezhednevnaia gazeta*, Jan. 18, 1995; Sergei Parkhomenko, "Spetssluzhbam ne udalos' vygnat' Dzhordzha Sorosa iz Rossii," *Segodnia*, Jan. 18, 1995.

26. Laura Belin, Open Media Research Institute (OMRI) report, Mar. 15, 1995.

27. Igor' Goriunov, "Gde ty, partiia nauki?" *Poisk*, Nov. 4–10, 1995, p. 1.

28. Michael Specter, "In Russia's Science City, Voting for Past Glory," *New York Times*, June 9, 1996, pp. 1, 8.

29. Grigory A. Yavlinsky, "Death of a Scientist," *New York Times*, Nov. 15, 1996, p. A33.

30. Vera Sandomirsky Dunham, *In Stalin's Time: Middleclass Values in Soviet Fiction* (1976; enl. and updated ed., with new introduction by Richard Sheldon, Durham, N.C. : Duke University Press, 1990).

31. Leonid Kosals, "Military R&D Institutes in the Context of Demilitarization in Russia" (working paper 94-002 [Jan. 1994], International Institute for Applied Systems Analysis, Laxenburg, Austria), p. 9.

32. Unpublished paper, St. Petersburg, 1995.

Chapter 3

1. The dean of Soviet airplane designers, A. N. Tupelov (whose name is still carried on the TU series of planes), worked for many years as a prisoner in a design laboratory established by the secret po-

lice. See G. Ozerov, *Tupolevskaia sharaga* (Frankfurt a/M: Possev, 1971). For the names cited in subsequent notes, I have drawn extensively on my own previous research plus a number of other studies, including M. G. Iaroshevskii, *Repressirovannaia nauka*, 2 vols. (St. Petersburg: Nauka, 1991, 1994); *Tragicheskie sud'by: repressirovannye uchenye akademii nauk SSSR; Sbornik statei*, ed. V. A. Kumanev (Moscow: Nauka, 1995); V. E. Boreiko, "Istoriia boli i geroizma: skorbnyi spisok deiatelei okhrany prirody i zapovednogo dela SSSR, repressirovannykh v 20–50-e gody," *Okhota i okhotnich'e khoziaistvo*, 1995, nos. 1–4; id., *Belye piatna istorii prirodookhrany: SSSR, Rossiia, Ukraina*, 2 vols. (Kiev: Kievskii ekologo-kul'turnyi tsentr, 1996); and David Joravsky, *The Lysenko Affair* (Cambridge, Mass.: Harvard University Press, 1970), which contains a list of "repressed specialists" (pp. 317–28).

2. Andrei Sakharov, probably the most famous scientist in Soviet history, was seized by the secret police in 1979 and held under house arrest in Gorky until released by Gorbachev in 1985.

3. Sergei Korolev, the "chief designer" of the Soviet space program and now the most famous scientist in Soviet space history, was arrested in 1937 and thrown into one of Stalin's labor camps, where he worked on rocket technology for many years in a special prison laboratory, or *sharashka*. He was rehabilitated after Stalin's death in 1953 and went on to develop military missiles that became the basis of the space program.

4. Lev Landau was arrested by the secret police in 1936 but was released on the pleading of Pyotr Kapitsa after a year in prison. Kapitsa himself was placed under house arrest for several years after he refused to participate in the atomic weapons project. Andrei Sakharov was placed under house arrest from 1979 to 1985.

5. Sergei Chetverikov, head of the genetics section of the Institute of Experimental Biology in biology, was a leader in the development of population genetics. He was arrested in 1929 and sent to exile in a remote Urals region. He was not rehabilitated until 1955, when he was too old to continue his work.

6. The most famous of all, of course, was Nikolai Vavilov, antipode to Lysenko, and a world-famous biologist. He was arrested in 1940 and died in prison of malnutrition in 1943. Vavilov's two suc-

cessors as president of the Lenin Academy of Agricultural Sciences, A. I. Muralov and G. K. Meister, were arrested and shot.

7. B. P. Gerasimovich, director of Pulkovo after 1933, was a famous astronomer who had worked from 1927 to 1929 at Harvard Observatory in the United States. He was arrested in 1936 or 1937, along with many other astronomers, and was shot on Nov. 30, 1937.

8. A. A. Vladimirov was arrested in 1930.

9. D. F. Egorov, who held these positions at the time of his arrest and exile in 1930, was the teacher of the famed N. N. Luzin as well as many other outstanding mathematicians. Egorov, still under detention, died in a hospital in 1931.

10. L. A. Zil'ber was arrested in 1930.

11. B. V. Numerov, director of the Leningrad Astronomical Institute, was arrested in 1936 and shot on Sept. 13, 1941.

12. G. D. Karpechenko, a young and brilliant geneticist, was arrested in 1941 and died in prison in 1942.

13. B. B. Polynov was arrested May 11, 1937, and released Mar. 27, 1939. He was elected a full member of the Academy of Sciences of the USSR in 1946.

14. D. D. Pletnev was shot on Sept. 8, 1941.

15. S. A. Efremov was arrested in 1928 and later shot in prison.

16. Boris Hessen (Gessen) delivered one of the most famous papers on the history of science ever written, "The Social and Economic Roots of Newton's *Principia*," at the Second International Congress of the History of Science in London in 1931. He was arrested in August 1936 and shot on Dec. 20, 1936. Six of the eight members of the Russian delegation to London in 1931 were arrested.

17. A. I. Postoev, director of the Tashkent Observatory, was arrested during the purges of 1936–37; he lived in the camps for a number of years, then escaped with the German army in World War II and ended up as a displaced person in the American zone of Germany at the end of the war. Threatened with forced repatriation to the USSR (and almost certain death), Postoev managed to get an invitation from the American astronomer Harlow Shapley to take a position at the Harvard College Observatory. However, the U.S. government denied him a visa on the basis that he might be a security risk. Thus Postoev faced the double tragedy of accusations in the So-

viet Union of being a "capitalist wrecker" and in the United States the suspicion of being a communist spy. Postoev emigrated in 1952 to Brazil, where he worked in an astronomical institute. He died in an automobile accident in 1977.

18. V. R. Berg was arrested in 1933 or 1934.

19. S. V. Korshun was arrested and shot in 1931.

20. Ia. N. Afanas'ev was arrested in 1937 and died in detention the following year.

21. I. V. Obreimov and A. I. Leipunskii were both arrested and subsequently released in the period 1937–38.

22. V. V. Stanchinskii, developer of trophic dynamics, was denounced and arrested in 1933 and spent several years in a prison kolkhoz run by the secret police. He was rearrested June 23, 1941, and died in prison in 1942 in Vologda.

23. N. P. Gorbunov, former personal secretary to Lenin, was rector of this institution from 1923 to 1929. He was arrested in 1938. Cited dates of his death vary greatly, but it appears most likely that he was shot on the day of the arrest, Sept. 7, 1938. The other rector of this engineering institution who was arrested was I. A. Kalinnikov, one of the defendants in the Industrial Party Trial of 1930.

24. V. V. Parin was arrested and spent seven years in prison.

25. L. S. Shtern was, before the Russian Revolution, the first woman to chair a department at the University of Geneva. She was arrested in Moscow on Jan. 28, 1949, and sent in prison exile to Kazakhstan. After Stalin's death in 1953, she was released and returned to Moscow.

26. G. E. Ermakov was arrested in 1937.

27. V. O. Mokhnach was arrested in 1937 and released from prison in 1956.

28. A. A. Ianata was arrested in 1933.

29. E. R. Gesse was arrested and shot in 1937 or 1938.

30. S. D. Shein was the head of the Inter-Bureau of Engineering Sections (VMBIT), a union organization open to both engineers and technicians, which had a membership in 1927 of 105,000. He was arrested and imprisoned in 1930.

31. V. A. Barykin was arrested in 1937.

32. Petr Pal'chinskii was arrested in April 1928 and shot in May 1929.

33. S. M. Nekrashevich was arrested in Dec., 1929.

34. S. G. Levit was the founder of this institute. He was arrested in 1936 or 1937.

35. The prominent political leader Nikolai Bukharin, who also served in this academic post, was shot in 1938.

36. A. M. Bykhovskaia was arrested in 1937.

37. I. Ia. Bashilov was arrested in 1937 and died in exile in Krasnoiarsk in 1953.

38. A. A. Nurinov was arrested in 1937.

39. I. F. Grigor'ev was arrested in 1949 and died under interrogation.

40. A. A. Balandin was arrested twice, first in 1936 and again 1949, and spent a total of seven years in detention (1937–39; 1949–53). In 1943, between prison terms, he was elected a full member of the Academy of Sciences of the USSR.

41. A. L. Chizhevskii was arrested in 1936.

42. I. L. Krichevskii was arrested in 1938.

43. Iu. Figatner, "Proverka apparata," *VARNITSO*, Feb. 1930, p. 73, and "Chistka apparata akademii nauk: pochemu nuzhna byla chistka," *Izvestiia*, Aug. 30, 1929, p. 4.

44. Robert A. McCutcheon, "The 1936–1937 Purge of Soviet Astronomers," *Slavic Review* 50 (Spring 1991): 100–117.

45. Charles Gillispie, "Stalin in the Laboratory," *New York Times Book Review,* Mar. 21, 1993, p. 25.

46. The chairman of the committee was Dr. Carl Kaysen, director of the Institute of Advanced Study in Princeton. The final published report was entitled *Review of U.S.–USSR Interacademy Exchanges and Relations* (Washington, D.C.: National Academy of Sciences, 1977).

47. Science under Hitler also suffered grievously, but the losses were neither as great nor extended over such a long time. See Paul R. Josephson, *Totalitarian Science and Technology* (Atlantic Highlands, N.J.: Humanities Press, 1996).

48. Irina Dezhina, "Nauka: Sostoianie sfery issledovanii i razrabotok v pervom polugodii 1995 g.," in *Rossiiskaia ekonomika v pervoi polovine 1995 goda: tendentsii i perspektivy* (vyp. 13) (Moscow: Institute of Economic Problems of the Transition Period, 1995), pp. 82–86.

49. Ibid.

50. See *International Science Foundation, 1994: Annual Report* (Washington, D.C., 1995).

51. Irina Dezhina, "ISF Activity in Russia (Long-Term Grants for Research): Regional and Structural Aspects" (unpublished paper given by author to Loren Graham, 1994), p. 1.

52. Dezhina, "Nauka: Sostoianie . . ."

53. *Zarubezhnaia pomoshch' nauke i vysshei shkole Rossii: Spravochnik* (Moscow: Kur'er RAN i vysshei shkoly, 1994).

54. Dezhina, "Nauka: Sostoianie . . ."

55. An example of such hyperbolic articles is Andrei Vaganov, "Akademicheskai nauka ne umirala; v Sovetskoi Rossee ee prosto ne bylo" (Science in the Academy has not died; in Soviet Russia it simply never existed), *Nezavisimaia gazeta*, Sept. 1993.

56. V. E. Zakharov and V. E. Fortov, "Poteriaem li my okonchatel'no fundamental'nuiu nauku v Rossii?" (copy of manuscript in personal possession of Loren Graham; an abbreviated form appeared in *Izvestiia*, Nov. 3, 1994).

57. See "Storm Clouds over Russian Science," *Science* 264 (May 27, 1994), pp. 1259–82.

58. Glenn E. Schweitzer, "Can Research and Development Recover in Russia" (unpublished paper supplied to Loren Graham), p. 20.

59. Irina Dezhina, "Adjustment of Russian Science and Brain Drain," STS Program, MIT (unpublished paper, 1997).

60. I visited these institutions as a member of the board of the International Science Foundation, an organization set up by the philanthropist George Soros to help Russian science.

61. Paul Josephson, *Physics and Politics in Revolutionary Russia* (Berkeley: University of California Press, 1991). And see also *Fizikotekhnicheskii institut imeni A. F. Ioffe* (Leningrad: Nauka, 1978).

62. Described by Andrei Sakharov in his *Memoirs*, trans. Richard Lourie (London: Hutchinson, 1990), pp. 113–14. It is true that espionage played an important role in the Soviet atomic bomb project, and secret information from the Manhattan Project undoubtedly helped the USSR build its atomic bomb. But most analysts agree that the absence of such information would have delayed the Soviet atomic bomb only by a few years at most. Furthermore, the USSR

developed the hydrogen bomb on its own, using an original design. It had a deliverable hydrogen bomb before the United States.

63. David Holloway, *Stalin and the Bomb: The Soviet Union and Atomic Energy, 1939–1956* (New Haven: Yale University Press, 1994).

64. Andrei Sakharov, *Vospominaniia* (New York: Chekhov, 1990), pp. 155–86.

65. David Holloway, the major historian of the Soviet atomic project, carefully documents the espionage activity of Klaus Fuchs, Alan Nunn May, David Greenglass, John Cairncross, and Donald Maclean. Furthermore, Holloway believes that other spies on the American side existed who have not been revealed to the present day. He states that "someone was passing on information about the work of Seaborg and Segré at Berkeley" and cites a Soviet KGB source as stating that "at most one half" of the agents in the Manhattan Project passing information on to the Soviet Union were ever uncovered. At the same time, Holloway rejects as baseless the sensational claim of the Soviet spy master Pavel Sudoplatov that such famous scientists as Oppenheimer, Fermi, Bohr, and Szilard knowingly passed on atomic secrets to the Soviet Union. See Holloway, *Stalin and the Bomb*, pp. 82–84, 90–95, 103–8, 138, 174, 222–23; and Pavel Sudoplatov and Anatoli Sudoplatov (with Jerrold L. and Leona P. Schecter), *Special Tasks: The Memoirs of an Unwanted Witness—A Soviet Spymaster* (Boston: Little, Brown, 1994), esp. pp. 3, 172, 192, 196–97.

66. See p. 158.

67. *Poisk*, 1995, no. 47: 7.

Chapter 4

1. The phrase "dramatic political interference in academia" is taken from the announcement mailed out to participants in the "Wissenschaft und Macht" forum held in Halle, Germany, May 15–17, 1996, where an earlier version of this chapter was presented. I would like to express my appreciation to Wolfgang Arnold and Manfred Heinemann for arranging that conference. A later version of the chapter was presented at the conference "Academia in Upheaval" in Trondheim, Norway, Apr. 3–7, 1997. I am grateful to Michael David-Fox and György Péteri for arranging that conference.

2. An enormous literature exists on these subjects. A few exam-

ples are: Kendall Bailes, *Technology and Society under Lenin and Stalin* (Princeton: Princeton University Press, 1978); Loren Graham, *The Soviet Academy of Sciences and the Communist Party, 1927–1932* (Princeton: Princeton University Press, 1967); David Joravsky, *The Lysenko Affair* (Cambridge, Mass: Harvard University Press, 1970); Zhores Medvedev, *The Rise and Fall of T D. Lysenko* (New York: Columbia University Press, 1969); Mark Azbel, *Refusenik: Trapped in the Soviet Union* (Boston: Houghton Mifflin, 1981); Mark Popovsky, *Manipulated Science* (Garden City, N.Y.: Doubleday, 1979); and *Repressirovannaia nauka*, ed. M. G. Iaroshevskii (Leningrad, 1991).

3. Loren R. Graham, *The Soviet Academy of Sciences and the Communist Party, 1927–1932* (Princeton: Princeton University Press, 1967); Aleksey E. Levin, "Expedient Catastrophe: A Reconsideration of the 1929 Crisis at the Soviet Academy of Science," *Slavic Review* 47, no. 2 (1988): 261–79.

4. James Andrews, "Science and the Public Sphere" (MS) and conversation and correspondence with the author.

5. M. Iu. Sorokina, "'Molchat' dalee nel'zia . . .' (Iz epistoliarnogo naslediia akademika S. F. Ol'denburga)," *Voprosy istorii estestvoznaniia i tekhnika*, 1995, no. 3: 110.

6. Alexander Vucinich, *Empire of Knowledge: The Academy of Sciences of the USSR (1917–1970)* (Berkeley: University of California Press, 1984).

7. Jeremy R. Azrael, *Managerial Power and Soviet Politics* (Cambridge, Mass.: Harvard University Press, 1966), pp. 29–30. Also see Kendall Bailes, *Technology and Society Under Lenin and Stalin* (Princeton: Princeton University Press, 1978).

8. Hilde Hardeman, *Coming to Terms with the Soviet Regime: The "Changing Signposts" Movement among the Russian Emigration in the Early 1920s* (DeKalb: Northern Illinois University Press, 1994).

9. Alexandra Kollontai and V. Pletnev were very critical of scientists and engineers. See the discussion "The Great Debate over Technical Specialists," in Loren R. Graham, *Science in Russia and the Soviet Union: A Short History* (Cambridge: Cambridge University Press, 1993), pp. 88–90.

10. The following section draws on my "Big Science in the Last Years of the Big Soviet Union," *OSIRIS*, 2d ser., no. 7 (1992): 49–71.

11. Dmitry I. Piskunov, "Soviet Fundamental Science: State,

TABLE FOR NOTE 12
Organization of Research Personnel in the Soviet Union, ca. 1990

University system	Academy of Sciences system	Industrial and defense system
State Committee of Higher and Secondary Education of USSR	Academy of Sciences of USSR	Industrial ministries; defense ministry
Union-republic ministries; committees of higher and secondary education	Siberian division of the Academy in Novosibirsk; other branches and filials	
Higher educational institutions (*vysshie uchebnye zavedeniia*); universities and colleges, including leading universities (Moscow, Leningrad): 770	Academies of sciences of union republics: 14 Academies of Agricultural sciences; Medical Sciences; Pedagogical Sciences; Engineering (new)	Industrial research institutes (*otraslevye instituty*);† closed military ("postbox") research institutes
600,000 researchers*	125,000	800,000
7% of R&D budget	6.5%	87%

*Includes faculty members as well as researchers.
†E.g., institutes of steam turbines, coal mining, electronics.

Problems, and Perspectives of Development" (unpublished manuscript, Analytical Center for Problems of Socio-Economy and Science-Technology Development of the Academy of Sciences of the USSR, Moscow, 1991), p. 2.

12. The table above is drawn in part from ibid., pp. 2–3.

13. See Mark Adams, "Research and the Russian University," in *The Academic Research Enterprise Within the Industrialized Nations: Comparative Perspectives: Report of a Symposium* (Washington, D.C.: National Academy Press, 1990), pp. 51–65.

14. Piskunov, "Soviet Fundamental Science," p. 2.

15. Alexei Zakharov, "The Democratic Opposition in the Process of the Creation of the Russian Academy of Sciences" (unpublished paper, Apr. 23, 1993), pp. 1–2.

16. Thane Gustafson, "Why Doesn't Soviet Science Do Better Than It Does?" in *Social Context of Soviet Science*, ed. Linda Lubrano

FIGURE FOR NOTE 17
Economic Development and Size of Research and Development Personnel in State-Socialist and Capitalist Countries, 1985

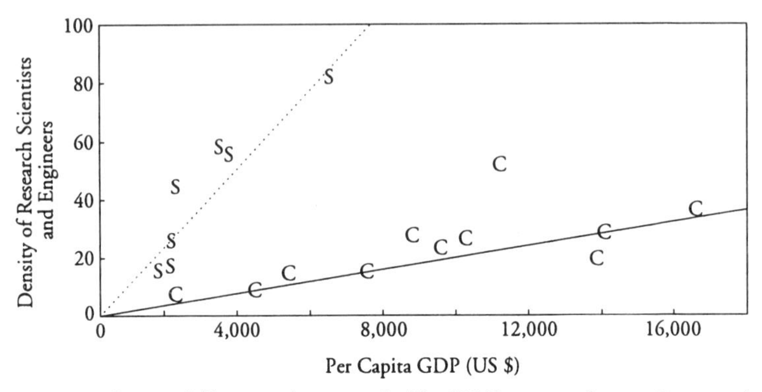

NOTE: State-socialist countries are marked by "S." From top: German Democratic Republic, Soviet Union, Bulgaria, Czechoslovakia, Hungary, Yugoslavia, and Poland. Capitalist countries, "C,": from right to left: the United States, Norway, Canada, Japan, Federal Republic of Germany, France, the Netherlands, Italy, Ireland, Spain, and Portugal.

SOURCES: Data for GDP from UN Department of Economic and Social Information and Policy Analysis, Statistical Division, *Statistical Yearbook, 1990–91* (New York: UN, 1993). Population data from the division's *Demographic Yearbook, 1991* (New York: UN, 1992). Data on workers in research and development in various countries are from UNESCO's *Statistical Yearbooks.*

and Susan Solomon (Boulder, Colo.: Westview Press, 1980); and *Post-izhenie: Sotsiologiia, sotsial'naia politika, ekonomicheskaia reforma*, ed. F. M. Borodkin et al. (Moscow: Progress, 1989), pp. 140–68.

17. The graph above is taken from György Péteri from p. 320 of his "On the Legacy of State Socialism in Academia," *Minerva* 33 (1995): 305–24.

18. Examples of outspoken views from this group were Maksim Frank-Kamenetskii, L. Tausen, and Boris Raushenbakh. Graham/Sloan Archive B2, p. 317; B2, p. 323; B2, p. 325.

19. V. D. Kazakov, "Science and Measures of Truth" (Feb. 5, 1989), Graham/Sloan Archive, B2, p. 318.

20. Perhaps the most outspoken and intelligent exponent of these views was Aleksei Zakharov from the Institute of Oceanography of

the Academy, and the leader of several different unions of scientific workers. See, e.g., his "Do We Need a Professional Union of the Academy of Sciences?" (Nov. 1, 1990), Graham/Sloan Archive 11.17, B8, p. 194.

21. V. Ginzburg, a full member, noted the resentments of the corresponding members, and favored an elimination of the dual ranks. However, he recognized the difficulty in achieving this change, and recommended, as a short-term easing of the problem, the granting of voting rights in the General Assembly of the Academy to corresponding members (V. Ginzburg, "Democracy in the Style of the Academy" [Apr. 14, 1990], Graham/Sloan Archive, 7.17, B2, p. 335).

22. The resentment of the engineers is discussed in K. Frolov, "A Russian Academy" (Dec. 29, 1989), Graham/Sloan Archive, 7.10, B2, p. 327.

23. L. Maiboroda represented this view ("Science Without a Ministry of Science" [May 18, 1991], Graham/Sloan Archive 7.55).

24. Resolution of the Supreme Soviet of the Russian Republic "On the Rules for the Formation and Organization of the Activity of the Academy of Sciences of the RSFSR" (July 13, 1990) (signed by Yeltsin), Graham/Sloan Archive 2.1.

25. V. Tishkov, "Is this the Academy or a Bastion?" Graham/Sloan Archive 7.20, B2, p. 339.

26. Gorbachev's decree was "On the Status of the Academy of Sciences of the USSR" (Aug. 23, 1990), Graham/Sloan Archive 2.2, B2, p. 116. The Supreme Soviet of the RSFSR canceled the part of Gorbachev's decree that applied to property in Russia: resolution of the Supreme Soviet of the RSFSR, "On Canceling the Validity of Article 2 of the Order of the President of the USSR 'On the Status of the Academy of Sciences of the USSR'" (Sept. 21, 1990), Graham/Sloan Archive 2.3. Yeltsin's cancelation of Gorbachev's decree in entirety is discussed in Andrey Allakhverdov, "Russian Scientists Gain Legal Rights . . . ," *Science* 273 (July 26, 1996), p. 424.

27. Iu. Danilin, "Academic Horrors" (Oct. 22, 1990), Graham/Sloan Archive 7.36, B2, p. 508; and G. Tolstikov, V. Kazakov, Iu. Manakov, V. Napalkov, and O. Chupakhin, "Their Own Academy Is Closer to Business?" (Feb. 27, 1991),Graham/Sloan Archive 7.49, B2, p. 530.

28. Quoted by Aleksei Zakharov, "The Democratic Opposition

in the Process of the Creation of the Russian Academy of Sciences" (unpublished paper, MIT workshop, Apr. 23–24, 1993), from *Nauka v Sibiri*, 1991, no. 2.

29. Zakharov, "Democratic Opposition . . . ," p. 38.

30. Ibid., p. 32.

31. As one Russian analyst observed: "In the lists of recipients of grants from the RFFI and the RGNF [Russian Humanities Foundation], one can find not only the directors of many institutes but also practically all members of the peer review boards and advisors to these foundations" (Irina Dezhina, "Nauka: Sostoianie sfery issledovanii i razrabotok v pervom polugodii 1995 g.," in *Rossiiskaia ekonomika v pervoi polovine 1995 goda: Tendentsii i perspektivy* [Moscow: Institute of Economic Problems of the Transition Period, 1995], pp. 82–86).

32. One of the most devastating criticisms was written by Academician Vitalii Ginzburg, who reveals just how corrupt the old system was. Ginzburg tells how the infamous Stalinist trial prosecutor Vyshinskii forced a system of election on the Academy that allowed observers to determine crucial votes. When voting for new members, the full members were given ballots on which there was only one candidate for each position, and there were columns after their names marked "elect" or "vote down." If a ballot was turned in "clean," that was considered a vote for "elect"; if a ballot was turned in with "vote down" alone underlined, that was a negative vote. What this meant was that any academician who pulled out his pen and marked the ballot was known to be voting against. All this was done in an open auditorium with the corresponding members and many other people present. This practice continued until almost the end of the Soviet period.

Ginzburg also describes the "telephone law" whereby Communist Party authorities and top leaders influenced procedures and decisions made by the Academy leadership. V. Ginzburg, "Democracy in the Style of the Academy" (Aug. 14, 1990), Graham/Sloan Archive 7.17, B2, p. 335.

Chapter 5

1. See *Does Technology Drive History? The Dilemma of Technological Determinism*, ed. Merritt Roe Smith and Leo Marx (Cambridge,

Mass.: MIT Press, 1994); Bruce Seely, "A Republic Bound Together," *Woodrow Wilson Quarterly* 17 (Winter 1993), pp. 19–39; and Todd Shallat, "Building Waterways, 1802–1861: Science and the United States Army in Early Public Works," *Technology and Culture* 31 (Jan. 1990): 18–33. Also see Thomas P. Hughes, *American Genesis: A Century of Invention and Technological Enthusiasm, 1870–1970* (New York: Viking, 1989).

2. Janet Ward, "100 Years of Public Works: 1894–1994," *American City and County* 109 (Sept. 1994): 74–95.

3. See Robert H. Wiebe, *The Segmented Society: An Introduction to the Meaning of America* (New York: Oxford University Press, 1975), p. 158; id., *The Search for Order, 1877–1920* (New York: Hill & Wang, 1967), p. 153; and id., *Businessmen and Reform: A Study of the Progressive Movement* (Cambridge, Mass.: Harvard University Press, 1962).

4. Thorstein Veblen, *The Engineers and the Price System* (New York: B. W. Huebsch, 1921).

5. See *Who's Who in America*, 1936–37; *Who's Who in Engineering*, 1937; and Hugh L. Cooper, "The Word 'Power'" (address delivered at commencement convocation, Univ. of Missouri, June 8, 1926).

6. Hugh L. Cooper, *Soviet Russia* (address delivered to the Institute of Politics, Williamstown, Mass., Aug. 1, 1930) (New York: Hugh L. Cooper, 1930), p. 8.

7. Michael J. McDonald and John Muldowny, *TVA and the Dispossessed: The Resettlement of Population in the Norris Dam Area* (Knoxville: University of Tennessee Press, 1982).

8. Cooper, *Soviet Russia*, p. 9.

9. Anne D. Rassweiler, *The Generation of Power: The History of Dneprostroi* (New York: Oxford University Press, 1988), pp. 45–56, 120–22.

10. I follow closely here the description I gave in my *The Ghost of the Executed Engineer: Technology and the Fall of the Soviet Union* (Cambridge, Mass.: Harvard University Press, 1993).

11. Rassweiler, *Generation of Power*, pp. 120–22.

12. Colonel Hugh L. Cooper, *American-Russian Trade During 1931 and Its Future Possibilities for the United States* (pamphlet from luncheon, Bankers' Club, New York, Jan. 28, 1932) (New York: American-Russian Chamber of Commerce, 1932), p. 8.

13. Stephen Kotkin, *Steeltown, USSR: Soviet Society in the Gorbachev Era* (Berkeley: University of California Press, 1991), pp. 208–28.

14. Hugh Cooper, *Trade with Russia* (pamphlet in Harvard Library) (n.p., 1931), p. 2.

15. Robert Conquest, *The Harvest of Sorrow: Soviet Collectivization and the Terror-Famine* (New York: Oxford University Press, 1987).

16. See Graham, *Ghost of the Executed Engineer*, esp. pp. 37–65.

17. The State Archive of the Russian Federation (GARF), f. 3348, op. 1, ed. khr. 639, l. 9–10; d. 562, l. 1; ed. khr. 41, l. 52–53.

18. GARF, f. 3348, op. 1, ed. khr. 751, l. 2.

19. Petr Pal'chinskii, "Gornaia ekonomika," *Poverkhnost' i nedra* 1, no. 29 (1927): 9.

20. Ibid.

21. Kotkin, *Steeltown, USSR*, p. 121.

22. Pal'chinskii, "Gornaia ekonomika," p. 9.

23. Kotkin, *Steeltown, USSR.*

24. Ibid., p. 254.

25. His personal papers are in the State Archive of the Russian Federation, Moscow, fond 3348.

26. Thomas P. M. Barnett, research paper, "Post-Stalinist Trends in the Soviet Politburo: The Development of Technocracy?" (Harvard University Government Department, Jan. 29, 1989).

27. A fact emphasized by Jonathan Spence, "A Flood of Troubles," *New York Times Magazine,* Jan. 5, 1997.

28. Kenneth Lieberthal and Michel Oksenberg, "The Three Gorges Dam Project," in their *Policy Making in China: Leaders, Structures, and Processes* (Princeton: Princeton University Press, 1988), pp. 269–338.

29. "China's Fickle Rivers: Rising Need for Water, and a Growing Crisis," *New York Times,* May 23, 1996, p. A6. Also Patrick E. Tyler, "Dam's Inexorable Future Spells Doom for Yangtze Valley's Rich Past," *New York Times*, Oct. 6, 1996, p. 12.

30. Patricia Adams, "Planning for Disaster: China's Three Gorges Dam," *Multinational Monitor,* Sept. 1993, p. 16.

31. Philip M. Fearnside, "Three Gorges Dam: 'Fatal' Project or

Step Toward Modernization?" *World Development* 16 (May 1988): 615–30.

32. Baruch Boxer, "China's Three Gorges Dam: Questions and Prospects," *China Quarterly*, no. 113 (Mar. 1988): 94–108. Also Mark A. McDowell, "Energy Strategies and Environmental Constraints in Chinese Modernization," *Chinese Geography and Environment* 3 (Fall 990): 2–23.

33. Adams, "Planning for Disaster"; Spence, "Flood of Troubles."

34. Wolfgang Bartke, *China's New Party Leadership: Biographies and Analyses* (Armonk, N.Y.: M. E. Sharpe, 1985); and Bartke, *Who's Who in the People's Republic of China*, 3d. ed. (Munich and New York: K. G. Saur, 1991).

35. Adams, "Planning for Disaster," pp. 16–20.

36. The English version, with new chapters describing events since its original publication in 1989, is Dai Quing, *Yangtze! Yangtze!* trans. Nancy Liu (Toronto: Earthscan, Canada, 1994).

37. Adams, "Planning for Disaster," p. 19.

38. Ibid.

39. Ibid., p. 19.

40. Audrey Ronning Topping, "Cracking the Wall of Silence," *New York Times Magazine*, Jan. 5, 1997, p. 40.

41. Adams, "Planning for Disaster," p. 20.

42. Patrick E. Tyler, "Cracks Show Early in China's Big Dam Project," *New York Times*, Jan. 15, 1996, pp. 1, A4.

43. The monthly subsidies were as low as $7.22 a month (ibid., p. A4).

44. Craig S. Smith, "China Dam Project Is Hard Sell Abroad," *Wall Street Journal*, May 3, 1995; "From Great Wall to Three Gorges," *Energy Economist*, Apr. 1992, pp. 6–8.

45. Tyler, "Dam's Inexorable Future . . . ," p. 12; and Spence, "Flood of Troubles . . ."

46. Tyler, "Dam's Inexorable Future . . . ," p. 12.

47. Ted Plafker, "China Says Yangtze Dam Will Be Safe," *Boston Globe*, Feb. 23, 1995, p. 16.

48. Karl Huus, "More Dam Trouble," *Far Eastern Economic Review*, Oct. 20, 1994, p. 72.

49. Quoted in ibid., p. 72.

50. David Luberoff, Alan Altshuler, and Christie Baxter, *Mega-Project: A Political History of Boston's Multibillion Dollar Artery/Tunnel Project* (Cambridge, Mass.: A. Alfred Taubman Center for State and Local Government, John F. Kennedy School of Government, Harvard University, June 1993; revised June 1994), p. 1.

51. Ibid., p. 2.

52. Ibid., p. 8.

53. Salvucci later described how his grandmother, an immigrant widow unable to speak English, was ordered to vacate her house in Brighton within ten days in order to make room for an urban highway. Interview of Thomas Parke Hughes with Fred Salvucci, MIT, Feb. 10, 1994, as reported in Hughes, "Coping With Complexity: Central Artery and Tunnel," in his *The Second Creation: Experts, Systems, and Computers* (New York: Pantheon Books, 1998).

54. Luberoff, Altshuler, and Baxter, *Mega-Project*, p. 11.

55. Kenneth Kruckemeyer, "Managing Public Participation in Public Works Projects" (Independent Activities Program Course, MIT, Jan., 1995; notes taken by Loren Graham).

56. Ibid.

57. Hugh G. J. Aitken, *Scientific Management in Action: Taylorism at Watertown Arsenal, 1908–1915* (Princeton: Princeton University Press, 1985), p. 46.

58. "Group Urges Boycott of China Dam Project," *Boston Globe*, Feb. 22, 1995, p. 4.

59. Richard L. Meehan, *Getting Sued, and Other Tales of the Engineering Life* (Cambridge, Mass.: MIT Press, 1981), p. 241.

60. Fred Salvucci, "Managing Public Participation in Public Works Projects" (Independent Activity Program, item 391, Jan. 1995). The course was actually given by Kenneth Kruckemeyer.

61. Kenneth Kruckemeyer, "Managing Public Participation . . ." (MIT, Jan. 1995; notes taken by Loren Graham).

Conclusion

1. Loren R. Graham, *Science, Philosophy, and Human Behavior in the Soviet Union* (New York: Columbia University Press, 1987); *Science in Russia and the Soviet Union: A Short History* (Cambridge, Mass.: Cambridge University Press, 1993), esp. pp. 99–120.

2. Ernst Mayr, *This Is Biology: The Science of the Living World* (Cambridge, Mass.: Harvard University Press, 1997), esp. pp. 16–23.

3. In 1988, there were in Russia 2,963,000 workers in "science and scientific services"; in the same year there were 1,368,000 workers in the "fuel and energy industries" and 1,327,000 in the "metallurgical complex," for a total of 2,695,000. See *Narodnoe khoziaistvo RSFSR v 1988 g.* (Moscow, 1989), pp. 33, 353.

Index

Other Books by the Author on Russian and Soviet Science

The Soviet Academy of Sciences and the Communist Party, 1927–1932. Princeton: Princeton University Press, 1967.

Science and Philosophy in the Soviet Union. New York: Knopf, 1972.

Review of US–USSR Interacademy Exchanges and Relations (Rapporteur). National Academy of Sciences Report, Washington, D.C., 1977.

Between Science and Values. New York: Columbia University Press, 1981.

Red Star: The First Bolshevik Science Utopia (edited with Richard Stites). Bloomington: Indiana University Press, 1984.

Science, Philosophy, and Human Behavior in the Soviet Union. New York: Columbia University Press, 1987.

Science and the Soviet Social Order (edited). Cambridge, Mass.: Harvard University Press, 1990.

Science in Russia and the Soviet Union: A Short History. Cambridge, Eng.: Cambridge University Press, 1993.

The Ghost of the Executed Engineer: Technology and the Fall of the Soviet Union. Cambridge, Mass.: Harvard University Press, 1993.

Library of Congress Cataloging-in-Publication Data

Graham, Loren R.
What Have We Learned About Science and Technology from the Russian Experience? / Loren R. Graham.
p. cm.
Includes bibliographical references and index.
ISBN 0-8047-2985-9 (cloth : alk. paper). —
ISBN 0-8047-3276-0 (pbk : alk. paper)
1. Science—Social aspects. 2. Technology—Social aspects. 3. Science—History—20th century. 4. Technology—History—20th century. 5. Science—Social aspects—Russia (Federation). 6. Constructivism (Philosophy). I. Title.

Q175.5.G73 1998
306.4'5'0974—dc21 97-41635
CIP

This book is printed on acid-free, recycled paper.

Original printing 1998
Last figure below indicates year of this printing:
07 06 05 04 03 02 01 00 99 98